# THE SOLAR SYSTEM

# THE OUTER PLANETS

# THE SOLAR SYSTEM

# THE OUTER PLANETS

SALEM PRESS
A Division of EBSCO Publishing
Ipswich, Massachusetts

GREY HOUSE PUBLISHING

Cover photo: Saturn, glowing stars, and a moon of Saturn. (© Ocean/Corbis)

*The Solar System: The Outer Planets,* 2013, published by Grey House Publishing, Inc., Amenia, NY, under exclusive license from EBSCO Publishing, Inc.

∞ The paper used in these volumes conforms to the American National Standard for Permanence of Paper for Printed Library Materials, Z39.48 1992 (R1997).

paperback ISBN: 978-1-4298-3796-5
ebook ISBN: 978-1-4298-3802-3

# CONTENTS

# CONTRIBUTORS

Reta Beebe

Raymond D. Benge, Jr.

Alvin K. Benson

John L. Berkley

Jennifer L. Campbell

Dennis Chamberland

Joseph Di Rienzi

David G. Fisher

Michael P. Fitzgerald

Dennis R. Flentge

George J. Flynn

David Godfrey

C. Alton Hassell

Paul A. Heckert

John P. Kenny

Narayanan M. Komerath

Kristine Larsen

V. L. Madhyastha

George R. Plitnik

Howard L. Poss

Clark G. Reynolds

J. Wayne Wooten

Clifton K. Yearley

# CALLISTO

**Categories:** The Jovian System; Natural Planetary Satellites

*Study of Callisto, Jupiter's outermost natural satellite, has led to insights into the formation of the solar system, the possibilities for extraterrestrial life, and the protection from comet impacts that Jupiter gives to the inner planets of the solar system.*

## Overview

Callisto is the outermost of the four major satellites of the "gas giant" planet Jupiter. It was discovered with one of the earliest telescopes by Galileo Galilei in 1610. Hence, it is often referred to as one of the Galilean satellites. Callisto is one of the largest satellites in the solar system, ranking third behind Jupiter's Ganymede and Saturn's Titan. With a diameter of 4,800 kilometers (2,985 miles), it is nearly the size of the planet Mercury. Callisto is also tidally locked to Jupiter, meaning that its "day" is the same length as its month, 16.82 Earth days. As a result, the same side of the satellite always faces Jupiter, just as the Moon always presents the same face toward Earth.

If the Galilean satellites had personalities, Callisto would be a frail old man. Unlike the young and vibrant Io, Callisto has neither volcanoes nor large mountains anywhere on its surface. In fact, its total lack of geological activity, both above and below the surface, means that its surface most likely resembles what the satellite looked like during its formation. This is at least partly due to the lack of tidal forces from nearby Jupiter. The lack of squeezing and pulling from Jupiter's gravity reduced the heat and energy within the satellite, leading to a relatively tranquil geology. This unique surface gives astronomers and geologists a glimpse of not only the primordial Jovian system but also the primordial solar system.

Callisto's surface is twice as bright as Earth's moon but still much darker than the surfaces of its Jovian siblings. The first few kilometers of the surface layer is primarily ice, with a darker material having leaked in at some point. Callisto's surface is uniformly covered in craters and is thought to be the most cratered satellite in the solar system. These impacts are the primary force that has shaped the planet, and sometimes great rings appear around the impact craters. The two largest features, Valhalla and Asgard, are respectively 3,000 kilometers (1,865 miles) and almost 1,600 kilometers (1,000 miles) in diameter. While impacts have been the primary force in shaping Callisto's surface, data from the Galileo space probe in the late 1990's showed that some minor erosion has occurred. This erosion is thought to be carbon dioxide sublimating through cracks in the surface ice.

Along with these large impact craters, there are numerous crater chains, or catenae. After the 1979 Voyager flybys, the catenae were thought to be the result of debris from asteroid impacts. This idea was called into question after the spectacular impact of Comet Shoemaker-Levy 9 into Jupiter's atmosphere during late May, 1994. This comet had come within a special distance from Jupiter, known as the Roche limit, and been broken up by the force

*Jupiter's pockmarked moon Callisto, as imaged from the Galileo spacecraft in 2001.* (NASA/JPL/DLR)

of gravity. What was once one large comet was now a series of fragments traveling in formation. This event gave credibility to the idea of comets colliding with planets and satellites and has helped to explain Callisto's pockmarked surface.

While the surface has given scientists relatively overt information about the satellite's past, Callisto's interior remains shrouded in mystery and conjecture. With a density of 1.86 grams/centimeter³, Callisto's density is the smallest of the major Jovian satellites. Scientists at the National Aeronautics and Space Administration (NASA) believe that Callisto is made up of roughly equal parts rock and ice, but the exact internal structure is unclear. Early observations led Galileo scientists to believe that Callisto is undifferentiated, meaning it has the same composition throughout.

Most rocky bodies in the solar system, such as Earth, have multiple layers that form during their creation. Molten materials tend to separate out, or differentiate, due to density. Within Earth, for instance, there is a dense core of iron and some nickel. Moving away from the core are different layers of decreasing density. Initial readings from Galileo showed that this process had not taken place in Callisto. Newer data, from subsequent flybys, do not directly contradict this hypothesis but have made planetary scientists less certain. Further evidence for an undifferentiated interior comes from data showing that Callisto also lacks its own magnetic field, suggesting a lack of a metallic core.

Curiously enough, however, Callisto does alter Jupiter's magnetic field within its vicinity. Because this perturbation in the field arises from increased conductivity within the planet, scientists speculate that a subsurface ocean may exist. Only an ocean with the salinity similar to Earth's oceans could explain the readings.

Callisto also has an extremely thin atmosphere composed primarily of carbon dioxide. With a pressure millions of times lower than Earth's, the atmosphere appeared, based on data from the Galileo flybys of 1998-1999, to have formed relatively recently. These data led scientists to believe that the atmosphere was no more than four years old and due to a combination of processes known as photoionization and magnetospheric sweeping. Photoionization takes place when ultraviolet rays (the same rays that cause sunburns) come in contact with individual carbon dioxide ($CO_2$) molecules; each $CO_2$ molecule ejects an electron, similar to the way a solar calculator generates current. Removal of an electron causes the molecule to become charged. Since charges interact

with magnetic fields, Jupiter's enormous magnetic field acts like a giant broom and sweeps these ionized particles away from Callisto. Left unchecked, this process would eventually cause Callisto's atmosphere to fade away.

If the atmosphere is not transient, the carbon dioxide gas must be replenished on a continual basis. The obvious source of $CO_2$ gas is Callisto's icy surface. This ice would have to be located in a region that is permanently shadowed, away from direct light and protected from ionization. It has also been suggested that much of the carbon dioxide that exists on the satellite's surface, as well as this tenuous atmosphere, comes from the comet impacts that Callisto has sustained.

## Knowledge Gained

The vast majority of Callisto data comes from the Voyager flybys of the late 1970's and the multiple flybys of the Galileo spacecraft during the late 1990's. Before that, the satellite was, at best, a foggy image in ground-based professional telescopes and a minuscule, but predictable, pinprick of light in backyard telescopes. Even Hubble Space Telescope images taken in October of 1995 showed a blurry surface. Only uncrewed space probes would produce the information needed to gain further understanding.

Both Voyagers 1 and 2, which took images on their way to the outer solar system, revealed a relatively dead world, battered by impact craters. Two decades later, Galileo returned to focus purely on the Jovian system. Its more sophisticated instruments offered higher-resolution imagery, magnetometric information, and spectroscopic information.

Galileo's most significant discovery about Callisto was the possibility of an underground ocean, similar to Earth's oceans. The discovery of water in the solar system is always a major event because it is thought to be an essential ingredient for life. Water was already thought to exist on nearby Europa, and great efforts were made to ensure that Galileo would not contaminate the surface. This included deliberately driving the probe into Jupiter's atmosphere at the conclusion of the mission. Water on Callisto was a much bigger surprise. Could Callisto now be added to the small, but growing, list of potentially fertile worlds within our solar system?

The possibility of a subsurface ocean arises from data on the local magnetic field around Callisto. Callisto does not possess an interior magnetic field but orbits well within the boundary of Jupiter's magnetic field. During multiple flybys, Galileo measured this magnetic field and detected fluxuations in its intensity. The local magnetic environment around Callisto is similar to an electromagnet. Whereas electromagnets have magnetic fields that are induced by the flow of electrons through a looped wire, Jupiter's magnetosphere does the opposite, capturing charged particles from the solar wind and creating electric currents in space. Galileo's instruments showed that this magnetic field was altered by increased conductivity from the satellite itself. While surface ice would not have any effect, the phenomenon could be explained by a subsurface ocean with a salinity level similar to Earth's oceans, conduction of current due to the presence of dissolved salts. This hypothesis is supported by the fact that similar data were taken at Europa, where planetary scientists are more confident that water exists below the surface.

More controversial is the continuing debate over Callisto's differentiation, or lack thereof. This controversy arose from data regarding Callisto's moment of inertia, a measurement of mass that indirectly comes from a body's rotation. This is the phenomenon that controls an ice skater's rotation, increasing it if the arms are brought close to the body and decreasing it when the arms are extended outward. Planetary scientists take this information one step further to determine the composition of a planet or satellite. A moment of inertia of 0.40 would mean that Callisto is totally undifferentiated. Data from multiple passes by Galileo showed a moment of just 0.38, within one standard deviation of theoretical uniformity. This debate is likely to continue for many years, until another spacecraft is sent. Regardless of the answer, the idea that Callisto is not as differentiated as Ganymede, a satellite similar in size and in distance from Jupiter, hints at an interesting beginning of the Jovian system. Answering the question of Callisto's interior will give scientists insight into planet and satellite formation.

While its innards will remain a mystery, Callisto's surface has helped astromoners understand more about comets, comet impacts, and Jupiter's role as protector of the solar system's inner planets (those between it and the Sun). Before the discovery of Comet Shoemaker-Levy 9, the idea of comets impacting planets was not universally accepted. Watching the comet slam into giant Jupiter, and the subsequent "bruises" it temporarily left behind, made the idea of cometary impacts more acceptable. Scientists also learned that it was Jupiter that caused the comet to split into fragments in the first place, leading many to believe that the gas giant has done this in the past. The fact that crater chains exist on the Jupiter-facing hemisphere of Callisto is evidence of past impacts and further evidence that Jupiter is the vacuum cleaner of the solar system, keeping the inner planets safe from dangerous debris.

Finally, studying Callisto may reveal much about the future of humankind, specifically the possibilities of colonizing the solar system. Project HOPE, or Human Outer Planet Exploration, is a futuristic concept mission put forth by NASA. Part of this exploration would include a crewed mission to Jupiter, with a landing on Callisto. Callisto is an optimal choice for a human landing for two reasons. The first is its icy surface, which would provide both a

*Taken in 1979 by one of the Voyager spacecraft from about 200,000 kilometers, this image of Callisto shows a multiple-ring basin.* (NASA/JPL)

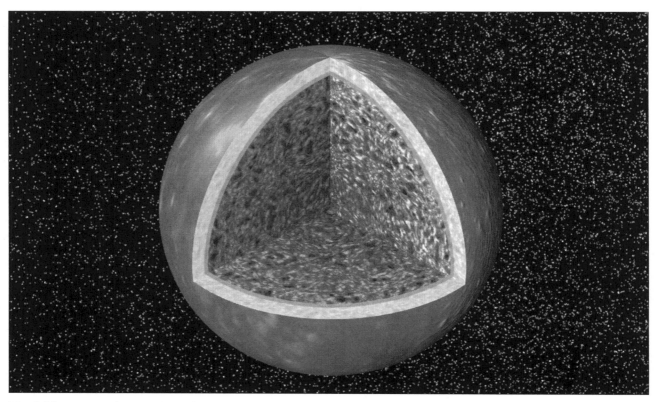

*The Galileo spacecraft returned data from Callisto that revealed that the Jovian moon may have a salty ocean underlying its icy crust, as shown in this artist's rendering.* (NASA/JPL)

source of water, allowing astronauts to "live off the land," and an opportunity for a first-rate study of impact geology. Second, Callisto's orbit places it in a region of low radiation from Jupiter. This remote, icy outpost would make an excellent location from which to study the Jovian system's past, present, and future.

## Context
Callisto is a wonderful example of how taking a second look leads to a different perception. The Voyager images offered snapshots of Callisto while racing through the solar system's highway. The Galileo probe effectively pulled over and took a look around. Missions like Galileo, which observed the Jovian system from late 1995 to 2003, and Cassini, which began observing Saturn in 2004, offer a chance to understand the distant gas giant planets along with their rocky satellites. Data from Galileo have pointed to the possibility of water on Callisto and have produced debates over its internal structure and its trace of atmosphere—all from a world

previously thought dead. Callisto has shown that every object in the solar system has a distinct and complicated personality, arising from a mysterious past, and that we have a long way to go when it comes to understanding our fellow travelers around the Sun.

*Michael P. Fitzgerald*

## Further Reading
Bagenal, Fran, Timothy E. Dowling, and William B. McKinnon, eds. *Jupiter: The Planet, Satellites, and Magnetosphere.* New York: Cambridge University Press, 2007. A collection of articles provided by recognized experts in their fields of study, this volume offers a comprehensive look at the biggest planet in the solar system. Excellent repository of photography, diagrams, and figures about the Jovian system and the various spacecraft missions that unveiled its secrets.

Carlson, Robert W. "A Tenuous Carbon Dioxide Atmosphere on Jupiter's Moon Callisto." *Science* (February

5, 1999): 283ff. A discussion of Galileo data regarding $CO_2$ in Callisto's atmosphere.

Cole, Michael D. *Galileo Spacecraft: Mission to Jupiter*. New York: Enslow, 1999. Provides a full description of the Galileo spacecraft, its mission objectives, and science returns through the primary mission. Particularly good at describing mission objectives and goals. Suitable for a younger audience.

Harland, David H. *Jupiter Odyssey: The Story of NASA's Galileo Mission*. New York: Springer Praxis, 2000. Provides virtually all of NASA's press releases and science updates during the first five years of the Galileo mission in a single volume, along with an enormous number of diagrams, tables, lists, and photographs. Also provides a preview of the Cassini mission. Although the book's coverage ends before completion of the Galileo mission, what is missing can easily be found on numerous NASA Web sites.

Khurana, K. K., et al. "Induced Magnetic Fields as Evidence for Subsurface Oceans in Europe and Callisto." *Nature* 395 (October 22, 1998). This article is the resource for all discussions of the possible subsurface ocean on Callisto.

Leutwyler, Kristin, and John R. Casani. *The Moons of Jupiter*. New York: W. W. Norton, 2003. Casani was the original Galileo program manager, and this book offers a heavily illustrated discussion of the Galilean satellites as well as a number of the lesser known Jovian satellites. The authors attempt to accompany their scientific findings with an artful text, which may please the tastes of some readers more than others.

McKinnon, William B. "Mystery of Callisto: Is It Undifferentiated?" *ICARUS* 130 (1997): 540-543. This article explains why the question of Callisto's differentiation does not have a definitive answer.

Melosh, H. J., and P. Schenk. "Split Comets and the Origin of Crater Chains on Ganymede and Callisto." *Nature* 365 (October 21, 1993). Discusses the hypothesis that crater chains on Callisto come from previous comets similar to Shoemaker-Levy 9.

Showman, Adam P., and Renu Malhotra. "Galilean Satellites." *Science* 286 (October 1, 1999). An excellent overview of Jupiter's four largest satellites.

# COMET SHOEMAKER-LEVY 9 COLLIDES WITH JUPITER

**Category:** Small Bodies

*When more than twenty fragments from Comet Shoemaker-Levy 9 collided with the upper atmosphere of Jupiter, the event offered scientists the opportunity to witness the consequences of the collision of extraterrestrial objects and provided insights into the likely effects of asteroid or comet impacts on Earth.*

## Overview

From July 16 to July 22, 1994, more than twenty fragments of Comet Shoemaker-Levy 9, which had been ripped apart by Jupiter's gravity during an earlier encounter with the planet, collided with the atmosphere of Jupiter. These collisions deposited more energy into the atmosphere of Jupiter than would be produced by all of the nuclear weapons in the military arsenals around the world. These impacts, which were observed from Earth and from spacecraft, provided astronomers with their first opportunity to witness a cosmic collision of a size capable of causing global consequences. The impacts produced bright fireballs in Jupiter's atmosphere and new cloud features visible from Earth.

Comet Shoemaker-Levy 9 was discovered by Eugene Merle Shoemaker, Carolyn Shoemaker, and David H. Levy in photographs taken on March 24, 1993. These photographs were taken using a 0.46-meter diameter Schmidt camera, a low-power telescope with a high light-gathering capability, at the Mount Palomar Observatory in Southern California. A Schmidt camera is designed so that it can see dim objects, and it has a wide field of view, making it an ideal instrument to search a large area of the sky for faint objects such as comets and asteroids. Shoemaker-Levy 9 was the ninth comet discovered by this group of researchers.

The comet was very dim—having a brightness of 13.8, much fainter than the dimmest object that can be seen with the human eye or with binoculars—when it was discovered. Still, it appeared unusual in the photographs, being slightly elongated. Once the position of Shoemaker-Levy 9 was determined, other observers looked at it with telescopes having higher magnification. Photographs taken by John Scotti, an astronomer using the Spacewatch telescope on Kitt Peak, Arizona, showed that Shoemaker-Levy 9 was not a single object but was actually several distinct objects spread out along the same path in space.

Astronomers referred to the comet as "a string of pearls" because its bright fragments were distributed in a line along its orbital path. Astronomers wondered what caused Shoemaker-Levy 9 to break into pieces. Other comets had been seen breaking up when they came close to the Sun, but the initial determination of the orbit of Shoemaker-Levy 9 showed that it had not passed close to the Sun.

Within days of its discovery, the comet had been observed by astronomers at the University of Hawaii and the McDonald Observatory in Texas. By April, 1993, these and other observations allowed Brian G. Marsden to determine that Shoemaker-Levy 9, instead of orbiting the Sun as is typical for comets, was actually in orbit around Jupiter.

Other researchers were able to trace the history of the comet's orbit. They determined that the comet passed only 15,500 miles above the clouds of Jupiter, within 1.4 Jupiter radii of the planet's center, on July 7, 1992. They suggested that during this close approach, the difference between the gravitational force Jupiter exerted on the near and far sides of the comet had ripped the weak comet into many pieces.

Shoemaker-Levy 9 had been in a rapidly changing orbit around Jupiter for several decades. The comet did not fragment during earlier encounters with Jupiter because it had approached no closer than about five million miles in its previous orbits. Analysis of high-resolution photographs taken by the National Aeronautics and Space Administration (NASA) Hubble Space Telescope in July, 1993, as well as images taken after the Hubble repair mission, which greatly improved the resolution, showed at least twenty-three discrete fragments, which were assigned letters A through W. The brightness of the Hubble images suggested that the visible fragments ranged in size from one-half mile to about one mile, with the fragments G and H being the largest. These visible fragments were embedded in a cloud of debris with material ranging from boulder-sized to microscopic particles.

By late May, 1993, it appeared that Shoemaker-Levy 9 was likely to hit Jupiter in 1994, and the fragments would be moving at a speed of about 130,000 miles per hour relative to the planet. At that point, the comet became the subject of intense study by astronomers around the world, since a cosmic collision of that magnitude had never been observed before. Based on the size and speed of each of the fragments, the impacts were expected to produce an explosive effect equivalent to between 6 million and 250 million megatons of TNT.

In the two-year period between the breakup of the comet and the collision with Jupiter, the fragments had spread out along the comet's orbit. The impacts, which took place over a one-week period from July 16 through July 22, 1994, caused enormous fireballs in the atmosphere of Jupiter and produced large, dark storms. Fragments G and H, each about 1.5 miles in diameter, caused the most destruction.

The first two impacts occurred on a part of Jupiter that was facing away from the Earth, so the impacts were not directly observable from Earth. However, astronomers were able to see the bright clouds of debris as they rose over the edge of the planet. An hour later, as Jupiter rotated, the impact sites became visible, and the extent of the

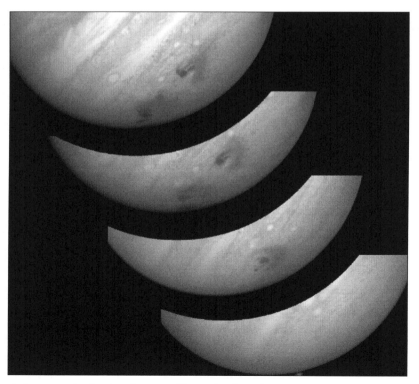

*A photo mosaic showing the comet Shoemaker-Levy 9 as it impacted Jupiter in 1994.* (JPL/NASA/STScI)

*The remains of Shoemaker-Levy 9 emerge after the comet's impact with Jupiter caused it to break into twenty-one pieces.* (JPL/D. Seal, edited by CXC/M. Weiss)

damage was clear. The impacts had left dark scars in the atmosphere of Jupiter.

The observation of Shoemaker-Levy 9's impact with Jupiter was a once-in-a lifetime event for astronomers. The disruption of a comet into many fragments is an unusual event. Capture of a comet into an orbit around Jupiter is even more unusual, and the collision of a large comet with a planet is extremely rare, estimated to occur only once in a thousand years.

## Significance

Sixty-five million years ago, Earth was struck by a large asteroid, an event which may have brought about the extinction of the dinosaurs. However, exactly what took place as the object passed through Earth's atmosphere has only been modeled, never verified, by experiment. The enormous release of energy into the atmosphere cannot be produced by humans, even with the use of nuclear bombs. Thus the Shoemaker-Levy 9 impacts into the atmosphere of Jupiter provided the first opportunity to observe this type of event and to validate the models, allowing scientists to determine

with better accuracy the likely effects of impacts of asteroids or comets on Earth.

Shoemaker-Levy 9's impact on Jupiter provided graphic visual evidence of the destructive power of comet and asteroid impacts on a planet. Governments around the world began to recognize the consequences such an event would have if it occurred on Earth. The event resulted in a more ambitious effort to discover and track asteroids and comets that approach Earth.

The impacts also allowed scientists to study Jupiter. The dark spots in the atmosphere were quickly distorted in shape, serving as a tracer to map the winds on Jupiter. Ultraviolet observations from the Hubble Space Telescope showed the debris sinking into Jupiter's atmosphere, providing the third dimension in the motion of Jupiter's winds. Observations also led to the suggestion that linear chains of craters, previously observed on two of Jupiter's moons, Ganymede and Callisto, might have formed by the impact of bodies disrupted in the same way as Shoemaker-Levy 9.

*George J. Flynn*

## Further Reading

Levy, David H. *Impact Jupiter: The Crash of Shoemaker-Levy 9*. New York: Basic Books, 2003. Well-illustrated, nontechnical account of the impact of Comet Shoemaker-Levy 9 into Jupiter. Discusses how the event changed the understanding of comets and cosmic cataclysms.

_____. *Shoemaker by Levy: The Man Who Made an Impact*. Princeton, N.J.: Princeton University Press, 2000. Account of the life of Eugene Shoemaker, including the events that led to the discovery of Shoemaker-Levy 9, and Shoemaker's work on the effects of comet and asteroid collisions with Earth and other heavenly bodies.

Noll, Keith, Harold A. Weaver, and Paul D. Feldman, eds. *The Collision of Comet Shoemaker-Levy 9 and Jupiter*. New York: Cambridge University Press, 2006. A 388-page collection of scientific reports presenting the major scientific results from observation of the impacts.

Shoemaker, Gene, Carolyn Shoemaker, John R. Spencer, and Jacqueline Mitton. *The Great Comet Crash: The Collision of Comet Shoemaker-Levy 9 and Jupiter*. New York: Cambridge University Press, 1995. Collection of images from telescopes around the world showing Comet Shoemaker-Levy 9 and the effects of its collision with Jupiter.

# ENCELADUS

**Categories:** Natural Planetary Satellites; The Saturnian System

*Enceladus is the brightest of the satellites located within the rings of Saturn. It has much in common with what scientists expected to find on comets, especially water ice. However, it appears to have been formed more than four billion years ago as a spinning mass of soft material.*

## Overview

Enceladus was discovered on August 28, 1789, by William Herschel. Its orbit around Saturn has a semimajor axis of 237,948 kilometers and an eccentricity of 0.0047, with a period of 118,386.82 seconds (nearly 33 hours). The orbit is inclined at 0.019° to Saturn's equator and located inside the E ring around Saturn. This places Enceladus at roughly 4 Saturn radii, located between the orbits of the moons Mimas and Tethys, which are about one Saturn radius on either side. Its rotation is synchronous and has no axis tilt, so that the same hemisphere always faces Saturn. It is nearly spherical, with a mean diameter of 504.2 kilometers, being a slightly flattened ellipsoid with dimensions of 513.2 kilometers along the orbit radius pointed at Saturn, 502.8 kilometers along the orbit path, and 496.6 kilometers between the north and south poles.

It is the sixth-largest moon of Saturn. It has a mass of $1.08022 \times 10^{20}$ kilograms, and the mean density is approximately 1,609 kilograms per cubic meter. The value of acceleration due to gravity at the surface on the equator is 0.111 meter/second$^2$. The escape velocity at the surface is 0.238 kilometer/second, neglecting atmospheric drag. Enceladus has very high Bond albedo of 0.99 and geometric albedo of 1.375, the highest among the satellites embedded in the Saturnian rings, indicating strong reflection. Its apparent magnitude from Earth is 11.7.

Enceladus is tidally locked in synchronous orbit around Saturn, meaning that the hemisphere with the higher density always faces Saturn. Looking into the sky from the Saturn-facing side, the planet would occupy roughly 30° of the sky and appear to be spinning in roughly the same position at all times. From the side hidden from Saturn, the Sun would appear very small, rising and setting in roughly 17 hours as Enceladus completed half an orbit around Saturn.

The orbital eccentricity of Enceladus is attributed to a resonance with the satellite Dione, with Enceladus completing two orbits for each orbit by Dione. This resonance may also drive the tidal heating of Enceladus. The shape of Enceladus is very close to that of an equilibrium-flattened ellipsoid, hydrostatically balanced by gravity and spin, which is the shape that an object would have in space if it were composed of homogeneous and fluid material. However, simulations of orbital evolution and the tidal locking suggest some variation in its internal density. The higher average density than that of water indicates denser material, possibly silicates, inside Enceladus.

Most of the data about the surface and orbital environment of Enceladus come from close flyby observations by the Voyager and Cassini spacecraft. The Voyager mission revealed evidence of a complex thermal history of Enceladus and showed several provinces with distinct geographical features. Short periods of intense heating and geological activity appear to have been separated by long periods of inactivity. Surface features include long, narrow depressions (fossae), ridges with cliffs of several

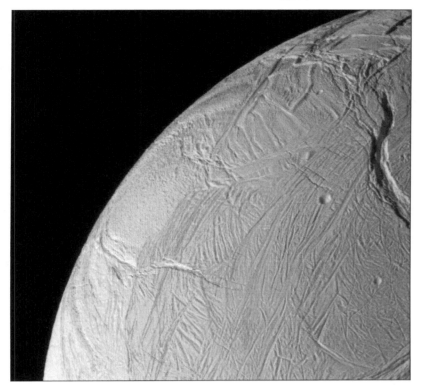

*The smooth, young surface of Enceladus is seen in this October, 2008, image from the Cassini spacecraft.* (NASA/JPL/Space Science Institute)

130 kilometers long, are flanked by 100-meter high ridges. Temperature is as high as 175 kelvins. South of 55° south latitude, a chain of fractures and ridges circumscribes the moon. Some fractures intersect and overlay others. Scientists associate these fractures and ridges with the flattening and extension due to gravitational interactions with Saturn and other moons.

The Cassini flyby of Enceladus in 2005 revealed a water-rich plume ejecting as narrow jets from vents in the tiger-stripe sulci of the south polar region. The fine sizes of the particles in the plume, which freeze to ice or sublimate soon after ejection, suggest that the plume originates in a subsurface body of liquid water. Some of the water jets reach exit speeds of 600 meters per second, well above the 238 meters per second needed to escape the gravity of Enceladus. The mass flow rate of the plume is on the order of 100 kilograms per second, comparable to that of air through a modern supersonic jet fighter engine. Much of this mass may initially escape Enceladus into the ring around Saturn and may be the origin of the mass in the E ring of Saturn. Enceladus also captures mass as snow falling to the surface from the E ring. Some scientists think that less than 1 percent of the mass in the plume can escape into the E ring. Numbers cited in different papers vary greatly on this subject.

Cassini approached to within 52 kilometers of the southern middle latitudes on March 12, 2008, in its E3 close approach, then dropped behind Enceladus in its orbit. It grazed the edge of the plume, which trails the satellite like the tail of a comet. The density of these portions of the plume had been predetermined from Earth by observing the dimming of starlight as the plume crossed in front of a star with an ultraviolet imaging spectrograph (UVIS). Roughly 70 seconds after closest approach, the craft was 250 kilometers above the surface, when the onboard ion and neutral mass spectrometer (INMS) encountered a peak particle density of nearly 10 million particles per cubic centimeter. This was still in the outer edge of the plume. The sharply defined plume edge provides further strong evidence that the plume comes out as a supersonic

hundred meters (dorsa), plains (plantia), long, parallel canyons (sulci), and craters. The terrain of the northern latitudes appears to be more than 4.2 billion years old, with more than one crater every 5 square kilometers, most of them bowl-shaped. Some craters are as large as 35 kilometers in diameter, while the majority are very small, less than a meter in diameter. The equatorial plains, named the Sarandib Plantia, show striations and folding, with about one crater every 70 square kilometers. Craters in these younger regions show viscous relaxation, indicating mechanisms for melting or distortion of the surface.

The ridged and grooved plains of the Samarkand Sulcus, at 55° to 65° south latitude, are 100,000 to 500,000 years old, and the fractured regions south of that show few if any craters. The Cassini orbiter approached the south polar region of Enceladus to within 168 kilometers on July 14, 2005, taking images with a resolution of 4 meters per pixel. House-sized boulders believed to be made of ice littered the polar landscape, but craters were nearly absent. A set of parallel "tiger-stripe" fractures, roughly 500 meters deep, 2 kilometers wide, and jet, as opposed to a diffuse subsonic plume caused by friction heating of ice due to the tidal stresses at the fractures. Measured gas density was twenty times higher than that predicted based on thermal expansion.

### Knowledge Gained

Data from the INMS suggest that the plume contains, beyond nearly pure water (ice), significant amounts (ranging from 1 to 10 percent) of carbon dioxide, methane, and other organic molecules, both simple and complex. There was a strong signal from something of molecular weight near 28, but there is debate over whether this substance is nitrogen (per the INMS data) or carbon monoxide (per the Cassini plasma spectrometer data). The values were similar to those predicted for many comet tails. The organic molecules detected include acetylene ($C_2H_2$), ethane ($C_2H_6$), hydrogen cyanide HCN, formaldehyde ($H_2CO$), propyne ($C_3H_4$), propane ($C_3H_8$), and acetonitrile ($C_2H_3N$). The cosmic dust analyzer instrument on the craft did not succeed in capturing the particle sizes. The Cassini plasma spectrometer (CAPS), which measures ions, detected much larger particles as well, on the order of nanograms. If these were ice particles, they may

have been as large as 0.01 to 0.1 millimeter in diameter, comparable to the particle sizes calculated from surface observation data of the south polar terrain. CAPS also detected positively and negatively charged ions, segregated in different regions of the plume.

Enceladus has no measurable internal magnetic field, but it has a significant influence on the magnetosphere of Saturn, strongly deflecting magnetic lines. This is now attributed to the water plume. Ions accelerated to energy levels of 20 kilo-electron volts by Saturn's strong magnetic field collide with the molecules in the plume, breaking them up into atoms and ionizing them. These fresh ions are again accelerated by the magnetic field, and, in turn, Enceladus substantially deflects the magnetic field lines of Saturn. Scientists also associate clouds of oxygen and hydrogen observed around Saturn with the atoms and ions generated when the water molecules in the plume from Enceladus collide with high-energy ions in the E ring.

Enceladus shows a trace atmosphere with surface pressure varying significantly in spatial location and perhaps in time, composed of about 90 percent water vapor, 4 percent nitrogen, 3.2 percent carbon dioxide, and 1.7 percent

*An artist's conception of the surface of Enceladus, showing one of the moon's ice geysers and Saturn in the background.* (NASA)

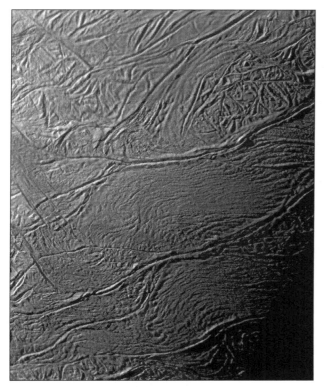

*Enceladus's "tiger stripes" are visible in this Cassini image from 2008.* (NASA/JPL)

methane. Surface temperature varies from 32.9 to 145 kelvins, with a mean of 75 kelvins. The surface appears to be covered in clean water ice, accounting for the high reflectivity. The atmosphere is hypothesized to be an expanding, supersonic neutral gas cloud emanating from the surface after molecules are "sputtered" from the surface by collisions with high-energy ions.

In late 2008, presentations given at the American Geophysical Union meeting in San Francisco revealed that data from Cassini's observations of Encedalus strongly indicated that the satellite's surface displays action similar to the action of Earth's ocean floor, where new crustal material emerges from slits in the crust. Cassini imaging team leader Carolyn Porco proposed that liquid water was present on Enceladus's surface and that that surface splits and spreads apart. On Earth, molten rock rising up from deep in the planet causes the crustal spreading, whereas on Enceladus the surface spreading originates with upwelling of liquid, presumably water. Evidence suggested that the Tiger Stripes formations near

the satellite's south pole are akin to the mid-ocean ridges found on the Earth's seafloor. Close flybys of that region resulted in more data on eruptions of water through vents in the Tiger Stripes formation. Combined with evidence of crustal spreading, these data have revealed Enceladus to be a surprisingly active world.

## Context

The presence of water ice in a low-gravity body within seven years' travel time of Earth excited planners of deep space missions, since water is an excellent future propellant. The discovery of high-speed water jets from the south polar region provides strong evidence of liquid water below the surface, and continuing tectonic processes. Complex organic molecules discovered in the jet plume fuel speculation about precursors of life. Enceladus is one of three known planetary bodies in the solar system (besides Earth and the Jovian satellite Io) that has an internal heat made visible by remote sensing. The relations between Enceladus and the E ring, the magnetosphere, and clouds of oxygen and hydrogen around Saturn are subjects of intense study.

The similarity between Enceladus and comets has raised questions about the origin and evolution of the solar system. Comets were thought to have originated far outside the orbit of Pluto, independently of the planets, while the planetary satellites formed from the same cloud as the Sun and planets. However, Enceladus appears to have nearly pure water ice, as predicted for comets. How such a large, nearly spherical cometary body could have been captured in an orbit so close to Saturn is an unanswered question.

*Narayanan M. Komerath and David G. Fisher*

## Further Reading

Benna, M., and W. Kasprzak. "Modeling of the Interaction of Enceladus with the Magnetosphere of Saturn." *Lunar and Planetary Science* 38 (2007). Discusses results of different numerical models of the magnetosphere interaction and the atmosphere of Enceladus, comparing against the results from the Cassini instruments.

Khurana, K. K., M. K. Dougherty, C. T. Russell, and J. S. Leisner. "Mass Loading of Saturn's Magnetosphere Near Enceladus." *Journal of Geophysical Research* 112, A08203, doi:10 .1029/2006JA012110, 2007. Reports on modeling of magnetic field data on the interaction between Saturn's magnetosphere and Enceladus. Gives results on mass pickup and current generated.

Porco, C. C., et al. "Cassini Encounters Enceladus: Background and the Discovery of a South Polar Hot Spot." *Science* 311 (March 10, 2006): 1401-1405. Discusses the initial discovery of the relatively warm regions near the south pole of Enceladus.

Verbiscer, A., R. French, M. Showalter, and Paul Helfenstein. "Enceladus: Cosmic Graffiti Artist Caught in the Act." *Science* 315 (February 9, 2007). Discusses the albedo of Enceladus compared to the albedos of other moons of Saturn.

Wilson, D., et al. "Cassini Observes the Active South Pole of Enceladus." *Science* 311 (March 10, 2006): 1393-1401. Presents images and discusses the reasoning regarding the features around the south pole of Enceladus, based on the Cassini spacecraft's close approaches in 2005. Also discusses how shape and size considerations are used to form hypotheses on the evolution of Enceladus.

# EUROPA

**Categories:** The Jovian System; Natural Planetary Satellites

*Europa is one of the four "Galilean satellites" that orbit the giant planet Jupiter. Only slightly smaller than the Earth's moon, Europa is covered by a relatively smooth layer of highly reflective fractured ice. Tidal forces exerted by Jupiter cause internal heating on Europa that apparently results in the periodic resurfacing of watery flows, which have over time obliterated most impact craters and other blemishes. Heat flow may be sufficient to maintain a liquid water subsurface layer that could harbor simple life-forms.*

### Overview

Europa is one of the four large satellites of the planet Jupiter known as the Galilean satellites after their discoverer, Galileo Galilei. These (according to their distance from Jupiter) are Io, Europa, Ganymede, and Callisto. Jupiter has at least sixty-three satellites, but only the Galilean satellites are large enough to be observed from Earth by small telescopes. With a diameter of 3,138 kilometers, Europa, the smallest Galilean satellite, is slightly smaller than Earth's Moon (3,476 kilometers). By contrast, the largest Galilean satellite, Ganymede, measures 5,260 kilometers in diameter, larger than the planet

Mercury (at 4,878 kilometers). Thus, if the Galilean satellites orbited the Sun instead of Jupiter, they would be considered full-fledged planets. Despite its relatively small size compared to its Galilean companions, Europa is nevertheless the sixth largest planetary satellite in the solar system. It is located about 780 million kilometers from the Sun, about 5.2 times the Earth-Sun distance.

Europa orbits Jupiter at an average distance of 670,900 kilometers; its orbital period (time to complete one orbit) is 3.55 Earth days. Its rotational period around its axis is also 3.55 days, which means that Europa always shows the same face toward Jupiter. The other Galilean satellites and Earth's own Moon follow this 1:1 ratio of orbital to rotational period, termed a "synchronous" relationship.

Galileo discovered Europa and two of the other three large Jovian satellites (Io and Callisto) on January 7, 1610, using a crude homemade telescope. At first he believed the tiny points of light in line with Jupiter were small stars, but later he realized that they in fact orbited Jupiter as if in a miniature solar system. Galileo originally called the moons the Medician planets (after the powerful Italian Medici family) and numbered each satellite with a Roman numeral beginning with the one closest to Jupiter. Europa in this scheme was designated II. Another observer, Simon Marius (Simon Mayr), who claimed to have discovered the Jovian satellites prior to Galileo in November, 1609, but was tardy in publishing his results, later named the bodies as we know them today.

The name Europa comes from a Phoenician princess, one of many mortal consorts of the supreme Greek god Zeus, whose Roman name graces the planet Jupiter. (The other Galilean satellites are similarly named for mythological characters associated with Zeus.) The most intriguing aspect of Europa is its unusual and unique surface. Images beamed to Earth in 1979 by the Voyagers 1 and 2 spacecraft as they flew through the Jupiter system showed a relatively smooth ice ball that some scientists compared in appearance to a fractured, antique ivory billiard ball. The satellite is covered by a globally encompassing shell of water ice, frozen at 128 kelvins, that gives Europa an extremely high albedo. While 64 percent of the light striking the surface is reflected back in all directions (giving Europa an albedo of 0.64), rocky surfaces like that of Earth's moon or Mercury reflect only about 10 percent.

Europa's density, 3.04 grams per cubic centimeter, suggests that most of the planet is composed of rocky silicate material like Earth. The icy surface layer, therefore, must be relatively thin; most estimates lie in the range of 75 to 100 kilometers thick. The surface shows relatively

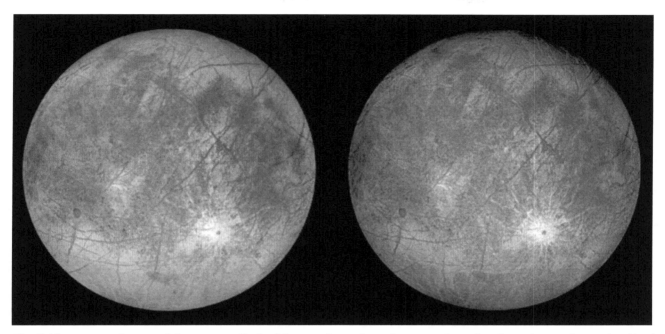

*The Galileo spacecraft returned these images (both of the same hemisphere, the one on the right enhanced to emphasize details) of Jupiter's satellite Europa, whose surface is primarily water ice. Darker areas are rocky material, lines are crustal fractures, and the bright spot on the lower right is the crater Pwyll. (NASA)*

little topographic relief, nothing higher than 1 kilometer, and displays only a few small scattered impact craters, in dramatic contrast to its highly cratered neighbors, the two outer satellites Ganymede and Callisto. Large craters on the order of 50 to 100 kilometers in diameter are virtually absent on Europa but plentiful on Ganymede and Callisto. Most craters on Europa do not exceed about 20 kilometers in diameter. This suggests that Europa's icy surface is relatively young, indicating that resurfacing by liquid ice flows or other processes has covered over any large craters formed during early, heavy meteoroid impacting in the Jovian system. Estimates of the surface age of Europa range from a high of 3.0 to 3.5 billion years old to more recent estimates of only 100 million years. The younger age, if true, suggests significant resurfacing of the planet in the later stages of its history.

In December, 1995, the Galileo spacecraft entered orbit about Jupiter. Over the course of thirty-eight orbits, Galileo not only investigated the giant planet Jupiter but also flew by its many satellites, paying particular attention to Europa, Ganymede, and Callisto. For example, in 1997 Galileo produced images of a large, multiringed impact crater on Europa probably buried beneath the ice crust. Evidence for the crater consists of diffuse, dark,

concentric, arclike bands and associated fractures that define a structure more than 5,000 kilometers in diameter. The presence of this buried crater shows that the rocky surface below the ice layer was subjected to significant impacting early in Europa's history. It further suggests that the ice crust formed at some later time, probably after heavy meteoroid bombardment had greatly diminished.

The most striking aspects of Europa's surface are the mottled, colored terrains and linear fractures that crisscross most of its globe. Mottled terrains, based on color and subtle topographic expressions, are of two varieties: brown and gray. Brown terrains contain numerous pits and depressions from 1 to 10 kilometers in diameter. Several large "plateaus" occur that range from a few kilometers up to a few tens of kilometers wide and up to nearly 100 kilometers long. Some circular depressions, missing raised crater rims, may represent degraded impact craters. Gray terrains are similar to the brown but are generally smoother and less hummocky. The relationship between the two terrains is unknown, but their differences may result from contrasting ages, degree of surface development, or both. The ultimate origin of these mottled terrains remains unknown, but a reasonable hypothesis is that they represent the effects of hydrothermal upwelling, causing heating and expansion of affected crustal areas. The "nonmottled" areas on Europa are very light in color and have very smooth topographies. These icy plains contain most of the observed linear surface fractures.

Linear features on Europa's surface may extend for thousands of kilometers. They are classified into three categories: (1) dark triple bands, some containing dark outer bands with a white strip down the center, thought to represent icy geyser deposits erupted along the axis of the fractures; (2) older and brighter lineaments that are crosscut by the triple bands and resemble them in some cases; and (3) very young cracks that crosscut the other two fracture types. Detailed analysis of the orientation of these three fracture types indicates that each type shows a distinct orientation that can be correlated with the relative age of the fractures. The data show that the direction of tidal stresses in Europa's crust has rotated in a clockwise direction over time. This observation has been used to suggest that Europa's rotation is not perfectly synchronous. Over time Europa may rotate faster than the synchronous rate, causing the surface to be progressively reoriented relative to tidal forces.

High-resolution images from the Galileo orbiter show places on Europa resembling ice flows in the Earth's polar regions. Large, angular pieces of ice have shifted away from one another, some rotating in the process, but reconstructions show that they fit together like puzzle pieces. This evidence for motion involving fluid flow, along with the possibility of geyser eruptions, shows that the ice crust has been, or is still, lubricated from below by warm ice, or even liquid water. The source of heating to produce this watery fluid is tidal forces by gravitational interaction with massive Jupiter, along with some escaping heat produced by radioactive minerals in the underlying silicate crust.

The Galileo spacecraft's mission was expanded to include the Europa Extended Mission, as it flew a number of close flybys to focus its instrumentation and cameras specifically on Europa. On one close encounter with Europa, Galileo came within 200 kilometers of the icy surface of the satellite. Ultimately, the Galileo spacecraft was purposely directed to plunge destructively into Jupiter's atmosphere on September 21, 2003. The reason for this was to safeguard any possible life-forms on Europa against

*An artist's cutaway showing the probable interior structure of Europa, with an iron core, a rocky mantle, and an outer ocean of salty water capped with ice.* (NASA/JPL)

the plutonium inside the Galileo spacecraft's radioisotope generators in the event that it might have crashed into the satellite.

Europa remains a high-priority location within the solar system for astrobiology studies. With the demise of the Galileo spacecraft, plans were proposed to send another spacecraft, this time to orbit Europa for prolonged and repeated studies. A more ambitious plan arose, called Jupiter Icy Moons Orbiter or JIMO. JIMO would have been the flagship mission of a more extensive program to develop nuclear propulsion as a means to cut down the time of travel between Earth and the rest of the solar system. That program was called Project Prometheus, but after initial funding was granted the National Aeronautics and Space Administration (NASA) was forced to postpone, if not cancel, this futuristic propulsion system in favor of other expenditures arising from the Vision for Space Exploration program under the George W. Bush administration. Returning to Europa with a robotic spacecraft was therefore put on hold for the early portion of the twenty-first century. Data from the Cassini spacecraft in orbit about Saturn revealed aspects about the icy satellite Enceladus and Titan (with its thick atmosphere and organic compounds) that diverted the attention of many astrobiologists away from Europa.

## Methods of Study

Jupiter and its four largest satellites have been studied using telescopes since Galileo first trained his on the system in 1610. Prior to the advent of interplanetary space probes, telescopic observations resulted in a remarkable treasure trove of data on the Galilean satellites.

For example, in the 1920's the astronomers Willem de Sitter and R. A. Sampson succeeded in obtaining reasonably accurate data on their masses. Calculations involved observing how each satellite disturbed the orbits of the others and by noting the nature of the resonant orbits of the inner three (first described by Pierre-Simon Laplace in the late eighteenth century). These resonant orbits dictate that for every one orbit of Io around Jupiter, Europa revolves two times and Ganymede four. This orbital resonance scheme implies a specific ratio for the masses of the bodies, which assisted de Sitter and Sampson in their calculations.

Diameters of the satellites were not accurately known until the advent of stellar occultation studies in the 1970's and later, when spacecraft imaging produced precise values. Prior to that, Europa was described by a popular 1950's-era science text as having a diameter of 1,800

miles (2,880 kilometers), only a bit less than the currently accepted value of 3,138 kilometers.

Although the first Earth-launched space probes encountered the Jovian system in 1973 (Pioneer 10) and 1974 (Pioneer 11), they paid scant attention to the Galilean satellites. The community of planetary scientists at the time viewed the satellites of the gas giant planets in the outer solar system to be nothing but rather boring ice balls. Even on the Voyagers, few planetary science studies were devoted to any of the icy satellites. Only imaging studies of Io and Titan, and to a lesser extent Europa, were planned as major portions of Voyager flyby operations in the Jupiter or Saturn system.

In 1979, however, knowledge of these bodies dramatically expanded as images of all four satellites were beamed back to Earth by Voyagers 1 and 2. The first pictures of Europa showed a previously unknown world, with a highly reflective, smooth surface mottled by brown and tan patches and crisscrossed by a complicated network of curved and straight lines. Four months later, higher-resolution images from Voyager 2 confirmed the presence of even more linear structures, which were interpreted as fractures but having virtually no relief associated with them. In addition to its imaging work, the Voyager probes also made precise measurements of the mass of the Galilean satellites by analyzing the gravitational effects of the planets on spacecraft trajectories, which, combined with improved size determinations, allowed for more accurate calculations of density. Density, in turn, is used to assess planetary composition.

In 1995 the Hubble Space Telescope discovered a thin oxygen atmosphere on Europa; this was later confirmed by the Galileo orbiter during its Europa Extended Mission. The official term for this rarefied atmosphere is a surface-bound exosphere. Hubble used its highly sensitive spectrometers to analyze the energy spectrum of light reflected from the moon's surface. Europa's atmosphere is so tenuous that its surface pressure is only one-hundred-billionth that of Earth. It is estimated that if all the oxygen on Europa were to be compressed to the surface pressure of the Earth's atmosphere, it would fill about a dozen Houston Astrodomes.

The Galileo spacecraft investigated the Jupiter system from late 1995 through 2003. After launching an atmospheric probe into Jupiter itself relatively early in its mission, the Galileo orbiter assumed an elliptical orbit that allowed it to make several close passes to all four Galilean satellites. The resolution of Voyager images of Europa made it possible to view surface features no smaller than

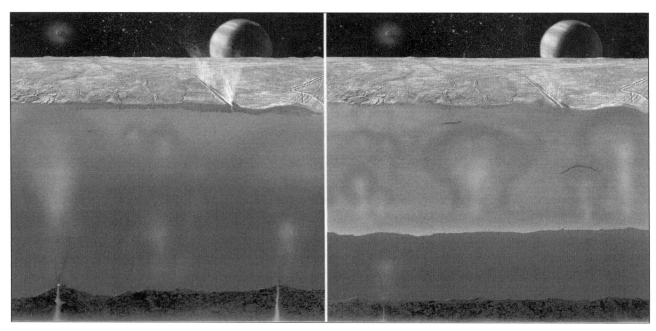

*An artist's cross-section rendering of two likely structures of Europa, with ice overlying a saltwater ocean beneath and heat rising through the rocky mantle, possibly volcanically.* (NASA/JPL)

about 4 kilometers across. In contrast, Galileo swooped down closer than either Voyager spacecraft, and with its more sophisticated cameras it achieved resolutions of around 10 meters per pixel, allowing objects the size of earthly buildings to be discerned. From these high-resolution images scientists have observed evidence of both tensional and compression ridges and have documented features like water-ice geysers, possible ice volcanoes, and jumbled ice flows that resemble puzzle pieces. These observations paint a picture of a dynamic planet in which tectonic faulting and flooding by liquid water occur periodically. The dark color of many surface fractures may result from the injection of water or warm ice mixed with darker silicates that well up into the fractures and freeze. Galileo images have generated renewed interest in the idea that a layer of liquid water exists below the ice or existed some time in the recent past.

Galileo also carried a magnetometer to detect the existence of a planet's magnetic field and to measure its strength. During a December, 1996, pass of Europa this magnetometer detected the first evidence of a magnetic field. Ganymede, the next moon out from Europa, also has a magnetic field. Although it is about four times weaker than that of Ganymede, Europa's field is still of substantial magnitude. Combined with gravity data suggesting a

dense core, the Europa magnetic measurements indicate the probable existence of a sizable metallic core, as well as a layered internal structure similar to that of Earth. The magnetic field data also, however, provided constraints on the nature of the water on Europa. The magnetic field could not be explained by assuming pockets of salty water within a crust of ice, but rather required a spherical shell of liquid water.

## Context

The Jovian system has long been of interest to scientists as a possible model analogous to the larger solar system of the Sun and planets. In this model Jupiter is a substitute for the Sun and the Galilean satellites represent the planets, particularly the rocky planets from Mercury to Mars. The considerable masses and stable orbits of the Galilean satellites suggest that they probably originated along with Jupiter during its formation from the gaseous solar nebula. If so, do these satellites show evidence for having evolved in a manner that parallels that of the inner planets of the solar system, including Earth?

In the early 1970's planetary scientist John Lewis pointed out that the densities of the two outer satellites, Ganymede (1.93 grams per cubic centimeter) and Callisto (1.83 grams per cubic centimeter), were consistent with condensation of solar-composition gas (the solar nebula), where water ice is a stable compound. He predicted that these two bodies should be composed of about equal parts water ice and silicate rock, a view generally accepted today. The two inner bodies, Io and Europa, however, have higher densities (3.55 and 3.04 grams per cubic centimeters, respectively) and thus would be expected to contain less in the way of low-density materials like ice, with a density of 1.0 gram per cubic centimeter. In fact, these bodies show evidence of being largely composed of rocky material, with no ice on Io and only a thin crust of ice over an ocean of liquid water on Europa. What processes could have produced such a density distribution?

In the early 1950's astronomer Gerard P. Kuiper suggested that Jupiter had been very hot during its early history. Building on Kuiper's early work, current hypotheses confirm that Jupiter was probably hot enough in its infancy to have forced low-mass, volatile gaseous materials to the outer fringes of the Jovian region, leaving heavier compounds to accrete as planetoids closer to Jupiter. The lighter volatile gas would contain a high proportion of elements that would eventually freeze as ice, compared to denser silicate minerals. These materials would eventually accrete to produce Ganymede and Callisto, while the volatile-poor inner gas would eventually accrete as Io and Europa. Europa, being farther from Jupiter than Io, has more volatiles such as ice than Io, which has virtually no ice on its surface. Io probably also lost much of its volatile component as a result of long-term volcanic activity, the result of tidal heating produced by Jupiter's gravitational field.

In a similar fashion the solar system as a whole exhibits a composition distribution with high-density "rocky" (or terrestrial) planets near the Sun and more volatile-rich bodies in the outer regions. The inner planets show a similar density distribution. Thus, the Jovian satellite system shows that any evolving planetary system on a scale large enough to have a hot central "star" predictably develops a density distribution where low-density, high-volatile planets dominate the outer regions and high-density "rocky" bodies dominate the inner regions.

The study of Europa is important in terms of its possible role as a site of extraterrestrial life-forms. Images from the Galileo space probe have confirmed ideas spawned after the Voyager flybys that Europa may have a globally encircling layer of liquid water beneath its surface ice layer. Where water exists in liquid form on a planet, life as we know it can theoretically evolve. Europa has joined the ranks of Mars and Saturn's moon Titan as possible sites where primitive life could exist.

Comparative planetology could indeed be actively studied inside the Jovian system by robotic spacecraft. Voyager provided only tantalizing images and raised many questions. Galileo data strongly suggested that the four large Jovian satellites share essentially similar cores in terms of size. Io is volcanically active, essentially taking material deep inside and turning the satellite inside out by resurfacing itself with sulfur and sulfur compounds; therefore Io is devoid of an icy shell. Europa has internal heat that is insufficient to melt the ice cover in total but supplies warmth to a subsurface layer of liquid water. Ganymede and Callisto have thick, icy shells and retain more of their primordial character because they lack significant heating from their cores. Ganymede's magnetic field is something of a paradox in that its character is akin to the dipole field produced by convection within a molten iron core.

Europa once held the primary attention of astrobiologists as the favored place within the solar system for finding some type of life beyond Earth. As a result of extensive robotic investigation of other portions of the outer solar system, it must now share that hopeful spotlight with Titan and Enceladus.

*John L. Berkley*

**Further Reading**

Bagenal, Fran, Timothy E. Dowling, and William B. McKinnon, eds. *Jupiter: The Planet,* Satellites, and Magnetosphere. Cambridge, England: Cambridge University Press, 2007. A comprehensive work about the biggest planet in the solar system, offering a series of articles by recognized experts in their fields of study. Excellent photographs, diagrams, and figures about the Jupiter system and the various spacecraft missions that unveiled its secrets.

Beatty, J. Kelly, Carolyn Collins Petersen, and Andrew Chaikin, eds. *The New Solar System.* 4th ed. Cambridge, Mass.: Sky, 1999. A chapter on the Galilean satellites by Terrence Johnson gives a comprehensive overview of these satellites based on Voyager and Galileo data. The volume is amply illustrated with color images, diagrams, and informative tables. Aimed at a popular audience, this book can also be useful to specialists. Contains an appendix with planetary data tables, a bibliography for each chapter, planetary maps (including Europa), and an index.

Cole, Michael D. *Galileo Spacecraft: Mission to Jupiter.* New York: Enslow, 1999. Provides a full description of the Galileo spacecraft, its mission objectives, and science returns through the primary mission. Particularly good at describing mission objectives and goals. Suitable for a younger audience.

Fischer, Daniel. *Mission Jupiter: The Spectacular Journey of the Galileo Spacecraft*. New York: Copernicus Books, 2001. Suitable for a wide range of audiences, this volume thoroughly explains all aspects of the science and engineering of the Galileo spacecraft. Particularly good discussions about the nature of the Galilean satellites.

Geissler, Paul E. "Volcanic Activity on Io During the Galileo Era." *Annual Review of Earth and Planetary Sciences* 31 (May, 2003): 175-211. The definitive work describing the physics and planetary geology of volcanoes on Io. Provides a complete picture of Voyager and Galileo spacecraft results.

Greely, R. *Planetary Landscapes*. 2d ed. Boston: Allen and Unwin, 1994. This book concentrates on the nature and origin of planetary surface features. It is packed with excellent diagrams, tables, maps, and monochrome images of planets taken by robotic spacecraft. A chapter on the Jovian system includes a detailed section on Europa, but it is dated prior to the arrival of the Galileo spacecraft into orbit about Jupiter. Contains an extensive reference section and index.

Harland, David H. *Jupiter Odyssey: The Story of NASA's Galileo Mission*. New York: Springer Praxis, 2000. Includes virtually all of NASA's press releases and science updates during the first five years of the Galileo mission, with an enormous number of diagrams, tables, lists, and photographs. Also provides a preview of the Cassini mission. Although the book was published before completion of the Galileo mission unfortunately, what is missing can easily be found on numerous NASA Web sites.

Hartmann, William K. *Moons and Planets*. 5th ed. Belmont, Calif.: Thomson Brooks/Cole, 2005. An updated version of a classic text on planetary science. The chapter on Jupiter covers all aspects of the Jovian system and spacecraft exploration of it.

Irwin, Patrick G. J. *Giant Planets of Our Solar System: An Introduction*. 2d ed. New York: Springer, 2006. Suitable as a textbook for upper-level college courses in planetary science. Focuses on Jupiter, Saturn, Uranus, and Neptune and their satellites, rings, and magnetic fields. Filled with figures and photographs.

Leutwyler, Kristin, and John R. Casani. *The Moons of Jupiter*. New York: W. W. Norton, 2003. Written by the original Galileo program manager, this heavily illustrated work provides discussions of the Galilean satellites and a number of the lesser known Jovian moons. The authors attempt an artful text to accompany the scientific findings, which may or may not be to the taste of all readers.

McBride, Neil, and Iain Gilmour, eds. *An Introduction to the Solar System*. Cambridge, England: Cambridge University Press, 2004. A complete description of solar-system astronomy suitable for an introductory college course. Filled with supplemental learning aids and solved student exercises. A Web site is available for educator support. Accessible to nonspecialists as well.

# GANYMEDE

**Categories:** The Jovian System; Natural Planetary Satellites

*The Jovian moon Ganymede was the first natural satellite, other than Earth's moon, to be discovered. It is also the largest satellite in the solar system—large enough to generate its own magnetic field, an unusual characteristic for a satellite.*

## Overview

Ganymede, the largest satellite of Jupiter, was discovered by Galileo Galilei with a telescope in 1610. He published the information in *Siderius Nuncius* (starry messenger) and thereby initiated a dispute with the Church that would eventually lead him to be placed under house arrest for the remainder of his life; it was heresy to say that anything revolved around something other than the Earth. The satellite was named by Simon Marius for one of the lovers of the Roman god Jupiter.

Ganymede is 5,280 kilometers in diameter, and just over 1 million kilometers from Jupiter; it is the seventh of sixteen satellites. It is common for a satellite to present the same face to its planet at all times—a relationship called synchronous—and Ganymede does this, just as Earth's moon does. The rotation of Ganymede is prograde, that is, in the same direction as that of Jupiter. Ganymede's orbit is almost circular, meaning that its eccentricity (the measure of how close to a circular orbit the satellite travels) is small. A circular orbit has an eccentricity of zero. Ganymede's angle of inclination is less than a degree, meaning that this moon rotates almost exactly in the plane of Jupiter's equator.

Ganymede's albedo, the amount of sunlight reflected, is large. This reflectivity is caused by ice mixed with carbon-rich soil on the surface of the satellite. When the ice underneath the surface is heated and melts, it erupts to the surface. The soil, which is denser than water, sinks below the water. The water freezes, causing a bright spot on the surface. The water is heated either by radioactive decay or by tidal flexing. Not only does the gravity of Jupiter and Callisto pull on Ganymede; the moon also has Laplace resonance, which occurs because of the forces from the satellites Io and Europa. Every time Ganymede rotates around Jupiter once, Europa, the satellite just inside Ganymede, goes around Jupiter twice, and Io, the moon inside Europa, goes around four times. Thus, during every orbit the three satellites are aligned, magnifying the gravitational effect. This increased gravitational pull and then relaxation not only causes the orbits to become elliptical but also causes stresses within the satellites themselves. This tidal flexing generates heat

that melts ice and causes the surface of Ganymede to be smoother than expected. Many of Ganymede's impact craters have had their depths reduced by the changing surface of Ganymede.

The surface of Ganymede is a mixture of bright areas and dark regions. The dark regions are heavily cratered; the bright areas are less cratered but are often grooved. The bright areas consist of the soil left after the ice has sublimed away. The surface appears much like the surface would if there were lava flow. Water melted by the tidal flexing, then freezing as it reaches the surface, may still flow, much as glaciers flow on Earth. The heavily cratered regions are older and often show fractures and grooves through the craters, where water has broken through the surface at a crater. Ice on Ganymede causes the reflection of radar to be much greater than on most other satellites. Ice also allows radar to penetrate more deeply into Ganymede than if the surface were all silicates. The percentage of ice has been measured at 45-55 percent.

The bulk density of Ganymede is between that of ice and that of carbonaceous silicates, indicating a mixture of the two materials. Ganymede has differentiated; that is, the components have separated, producing a core of dense metals, probably iron or iron with sulfur. A molten iron core is usually the reason for a magnetic field, which Ganymede does have. One model, which agrees with the measured values, postulates a large core of iron with 10 percent sulfur. The moon has a core with a radius of 695 kilometers, a silicate mantle, and a 900-kilometer-thick ice-water shell. Bombardment by meteors causes a change in the albedo of Ganymede in two ways. First, the meteor causes new, darker material to be thrown up onto the surface (the underlying silicates are darker than the ice or residue left by subliming ice). Second, meteor impacts cause the loss of volatile material, leaving an opaque, dark material.

Ganymede has a thin atmosphere, composed of electrically charged gases. One gas seen in the space around Ganymede is hydrogen. Water sublimed from the surface or escaping from

*This image of Jupiter's moon Ganymede shows the seismic activity in the crust and outer coating of water ice that has led to this body's nickname, the Moonquake World. It may be that Ganymede, like Earth, has plate tectonic activity.* (NASA, Voyager, © Calvin J. Hamilton)

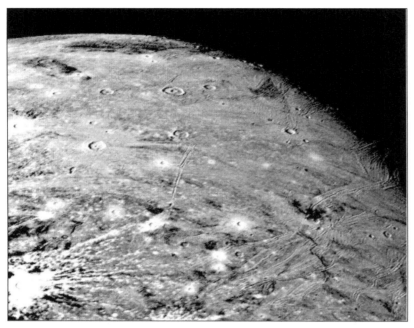

*Voyager took this picture of Ganymede during its 1979 flyby, from about 250,000 kilometers away. Impact craters, icy materials radiating out from them, ridges, grooves, and other surface features are visible down to about a 5-kilometer resolution.* (NASA/JPL)

Since Ganymede's orbit is in the same plane as Jupiter, it is thought that they were formed by the same process. Ganymede is out away from the very hot, dense region where the planet formed. Ganymede was formed in a cooler region, where water did not boil away but instead froze to form part of the satellite. Although Ganymede is locked into the same face toward Jupiter all the time, there are indications that this may not have always been true. One clue is that the number of meteor craters should be greater on the leading side of Ganymede, as is the case with Callisto, but this is not true of Ganymede. Another fact pointing to a change in the part of the ice shell facing Jupiter is the catenae that are found on the back side of Ganymede. Catenae are caused by a string of fragments from a comet that was broken up by the intense magnetic field of Jupiter but escaped capture to hit one of the satellites. They should occur only on the Jupiter-facing side of Ganymede.

a surface fracture condenses at the poles, producing a whitish polar cap down to latitudes of about 40°. Near-infrared spectra show the expected water and hydrated minerals. Unexpected are the indications of carbon dioxide, carbon bonded to hydrogen, carbon triple-bonded to nitrogen, sulfur bonded to hydrogen, and sulfur dioxide. The carbon dioxide appears to be trapped in the surface, perhaps in small bubbles. Jupiter's magnetic field causes ions to be swept along the orbit of the satellite, generating a current producing an auroral spotlight onto the poles of Jupiter.

Ganymede has an intrinsic magnetic field that is opposite to the field of Jupiter. It also displays an induced magnetic field caused by the strong rotating, angled field of Jupiter. The induced field is an indication of a conducting ocean deep under the icy surface. If the ocean has enough minerals dissolved in it to make it strongly conducting, it could generate the intrinsic magnetic field. Jupiter's strong magnetic field causes Ganymede to be bombarded by charged particles. This bombardment is thought to cause the molecular oxygen, $O_2$, and ozone, $O_3$, found in the surface of Ganymede.

## Knowledge Gained

The Jovian system, as Jupiter and its moons are called, has been visited by several space missions. Pioneer 10 (1973), Pioneer 11 (1974), Voyager 1 and Voyager 2 (both in 1979), and New Horizons (2007) flew through the system. They all used gravitational assists to gain speed to travel on toward the outer part of the solar system.

The Pioneer spacecraft provided the first visual images of the surface of Ganymede, as well as a much better estimate of the size and mass of Ganymede. The Voyager spacecraft improved these images to a resolution of a kilometer and provided color by means of six filters. Scientists were able to develop ideas of how the satellite formed and its structure. The Voyager data on craters indicated either that Ganymede's surface has changed, and thus erased the early craters from meteor hits, or that the surface was not firm enough to retain the early craters.

The Galileo mission arrived in orbit about Jupiter in 1996. The visual camera increased the resolution of the surface to 20 meters. The model of the moon, showing its core, mantle, and shell of ice, was developed after data from Galileo provided Ganymede's mass, average density, and moment of inertia. The moment of inertia and average density required the data on gravitational fields produced by Galileo, and how the flight path was perturbed as the craft flew by the satellite. Galileo also provided information of the magnetic fields using a magnetometer. The dual purpose of the magnetometer was to determine if Ganymede had a magnetic field of its own and how the satellites interacted with the strong magnetic field of Jupiter. Galileo used its Near Infrared Mapping Spectrometer (NIMS) to make a compositional map of the surface.

While studying Jupiter was not the main focus of New Horizons, astronomers did not let the opportunity escape. Although as of 2008 only 70 percent of the data had been transmitted to Earth, and only part of those data had been analyzed, New Horizons added some interesting new information. The spacecraft's infrared Linear Etalon Imaging Spectral Array (LEISA) and its panchromatic Long-Range Reconnaissance Imager (LORRI) charge-coupled device camera mapped Ganymede's composition. These instruments' resolutions are better than any land-based instrument or Galileo's NIMS. Low-temperature crystalline ice was found as expected, but asymmetric bands of non-ice were found, especially in the darker regions. More ice is found in bright regions and in craters and ejecta from recent meteor hits. This correlates with darker material on the surface, except where meteor strikes have brought ice to the surface. New Horizons could also map parts of Ganymede that Galileo could not see.

Not all information is gathered by spacecraft. The Hubble Space Telescope (HST) has taken pictures of the auroras of Jupiter. Other types of data, such as those gathered by eclipse radiometry, can be used from Hubble or from Earth. Eclipse radiometry is the measurement of thermal radiation just as the satellite is eclipsed by the planet. For Ganymede, these studies suggest that heat is lost rapidly; therefore, the surface material must be porous, due to bombardment from meteors over millions of years.

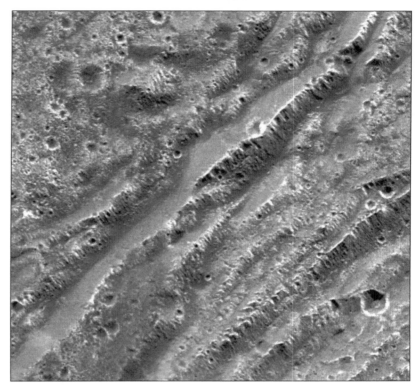

*The Galileo spacecraft flew by Ganymede more than two decades after Voyager, capturing this image of dark, heavily cratered terrain and scarps down to a resolution of about 20 meters.* (NASA/JPL/Brown University)

**Context**

The more astronomers learn about large bodies like Ganymede, the more is revealed about how the solar system was formed and about Earth and its Moon. Ganymede may be showing the action of plate tectonics. Learning about the plate tectonics of Ganymede may explain what happened on Earth as the continents tectonically rearranged.

Each space mission has returned valuable information on how to survive in space. Not only are meteorites a danger but gravitational wells, and especially strong magnetic fields, can damage a spacecraft. When humans venture forth, all of those dangers will have to be considered. The number of craters on Ganymede gives scientists some indication of the chance of a meteor hitting the Earth.

*C. Alton Hassell*

**Further Reading**

Asimov, Isaac, and Richard Hantula. *Jupiter*. Milwaukee, Wis.: Gareth Stevens, 2002. The famous science-fiction author covers the planet and its satellites, the space missions that have studied them, and the comet collisions of 1994. Illustrations, bibliography, index.

Corfield, Richard. *Lives of the Planets*. New York: Basic Books, 2007. The author takes the reader through the different space missions. Divided by planets, the information gathered by each mission is discussed. Index.

Fischer, Daniel. *Mission Jupiter: The Spectacular Journey of the Galileo Spacecraft*. New York: Copernicus Books, 2001. The author takes the reader through each step of the journey of this amazing space probe. Illustrations.

Grundy, W. M., et al. "New Horizons Mapping of Europa and Ganymede." *Science* 318 (2007): 234. This article is one of the first published after the flyby of Jupiter by the Pluto-bound New Horizons spacecraft. Illustrations, bibliography.

Leutwyler, Kristin. *The Moons of Jupiter*. New York: W. W. Norton, 2003. Ganymede is covered in one of the main sections of this book. Illustrations, index.

McFadden, Lucy-Ann Adams, Paul Robest Weissman, and T. V. Johnson, eds. *Encyclopedia of the Solar System*. San Diego: Academic Press, 2007. The editors have collected articles written by many experts in one of the best scholarly surveys of material about the solar system. Illustrations, appendix, index.

Slade, Suzanne. *A Look at Jupiter*. New York: PowerKids Press, 2008. Written for the juvenile audience, this book covers the important information in an easy-to-read style. One big section is devoted to Ganymede and Callisto. Illustrations, bibliography, index.

# IAPETUS

**Categories:** Natural Planetary Satellites; The Saturnian System

*The Saturnian satellite Iapetus is one of the most unusual satellites in the solar system. It has a dark side and a bright side, as well as a ridge along its equator that sits atop an equatorial bulge.*

## Overview

Iapetus was first noticed on one side of Saturn by Giovanni Domenico Cassini in October, 1671. Searching for it later on the other side of the planet was futile with the telescope that he was using. He tracked it many times over several years, but only when he had a better telescope in 1705 did Cassini finally see Iapetus on the other side of the planet. He concluded that Iapetus had a dark side and a bright side. It was named for the Titan god Iapetus, a brother of Cronus (the Greek name for the Roman god Saturn) in Greek mythology. Iapetus was originally called Saturn V, because the Saturnian satellites were originally numbered, and many scientists continued to use the numbers. With the discovery of other Saturnian satellites, Iapetus's scientific referent changed to Saturn VII and eventually Saturn VIII. It is one of the sixty satellites of Saturn that had been discovered by 2008. Iapetus's geological features, except for its dark region, are named after characters and places from the French epic poem *The Song of Roland*. The dark region is called Cassini Regio, in honor of Iapetus's discoverer, Cassini.

As Cassini deduced, Iapetus always presents the same face to Saturn, meaning it is synchronous with Saturn. Therefore the time of revolution and the time to circle the planet are identical, 79.32 days. Iapetus is prograde, meaning it turns in the same direction as Saturn. Iapetus's orbit is about 15.5° out of the plane of Saturn's equator; that is, its inclination is 15.5°. The inclination is large enough to cause questions about whether Iapetus was "captured" by Saturn's gravitational field or was generated by the same process that produced Saturn and the other satellites. Iapetus's elliptical orbit (its eccentricity is

*The Cassini spacecraft captured this image of Iapetus in September, 2007.* (NASA/JPL/Space Science Institute)

0.029) and a synchronous satellite that is 3,561,000 kilometers from the planet also add to the doubt about whether Iapetus was formed by the same process that formed Saturn.

Iapetus is an oblate spheroid with a 1,494-kilometer diameter along the axis pointed at Uranus. The equatorial axis is 1,498 kilometers, and the pole-to-pole axis is 1,426 kilometers. With its equatorial bulge and squashed poles, Iapetus has been said to look like a walnut. A 20-kilometer high ridge that girds most of the satellite along its equator is another striking feature. No other known satellite has a ridge like the one on Iapetus. The ridge is triangular, with a base that is about 200 kilometers wide. The ridge is also cratered, proving it has been in existence for a long time. One theory about the formation of the ridge is that the crust formed while the interior was still flexible enough for the weight of the shell to crush the interior. The interior material forced its way to the surface, fracturing the shell at the equator and forming a ridge. The material cooled relatively quickly, locking it in the new shape that is Iapetus today. This theory requires that Iapetus had to spin much faster during its formation and cooling periods. The slowing of its spin, called despinning, had to take place near a large object such as a planet. This, along with the fact that Iapetus's composition is similar to that of the other Saturnian satellites, indicates that Iapetus was formed in the process that formed Saturn and the other large satellites.

The first feature of Iapetus that astronomers noted was its albedo, which is completely different on the two sides of this satellite. The pictures from the Cassini spacecraft (2007) show a tar-black leading hemisphere with a bright backside hemisphere. The albedo is about 0.05 for the leading side and about 0.6 for the trailing hemisphere. This variation in albedo is noted not only in the visible range but also in the ultraviolet and radio ranges. The Cassini images indicated that the dark material is not solid but is in streaks large enough to appear solid from a distance. Near-infrared spectra indicate that the bright material is ice. The moon's density of 1.1 grams per centimeter cubed indicates that the rock fraction can be no more than 22 percent. The composition of the dark side appears to be ice contaminated with materials such as ammonia, amorphous carbon, poly-hydrogen cyanide (poly-HCN), and hematite ($Fe_2O_3$).

Some scientists have theorized that the dark material originates from another satellite, Phoebe. Phoebe has a retrograde orbit, meaning that it is traveling in the direction opposite to that of Saturn's rotation and thus in the direction opposite to Iapetus's orbital motion. Material lost by Phoebe because of its retrograde motion can bombard the front face of Iapetus. However, data from the ultraviolet spectra collected by Cassini show that the composition of Phoebe is not the same as the composition of the dark material on Iapetus. There is the possibility that the dark material might be from Hyperion, since the composition of Hyperion does match that of Iapetus. A second option is that the material from Phoebe changes before it bombards Iapetus. The dark material is thought to be a thin layer containing only a small amount (5 percent) of ice. A third opinion is that the dark material is consistent with an external impact. This would cause the poles of Iapetus to be bright, which they are. The bright poles may be bright because of frost. The peak temperature of the dark side of Iapetus was measured at 130 kelvins, warm enough to allow ice to sublime and refreeze at the poles, producing bright areas. The dark material has a reddish color that might indicate organic compounds. Organics would darken with exposure to

*A landslide is seen in the Cassini Regio portion of Iapetus, imaged by the Cassini orbiter in 2004.* (NASA/JPL/Space Science Institute)

23

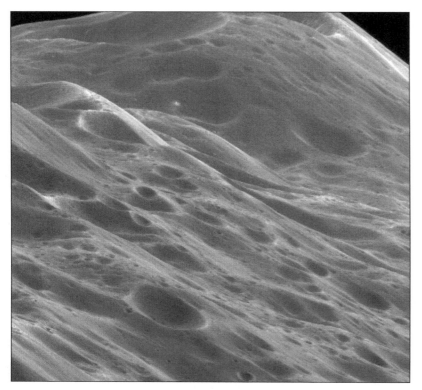

*Iapetus's mountainous terrain rises about 10 kilometers above the surface in this Cassini spacecraft image of the moon's equatorial ridge, taken on September 10, 2007.* (NASA/JPL/Space Science Institute)

radiation. Indications are that the material is porous and in the form of fine particles.

The topography of Iapetus was not well known before the Cassini mission. Even pictures from the Voyager 1 mission did not show any feature on the dark side. The bright side was also largely unknown. Iapetus is cratered, but is not saturated with craters. Around the craters were tall, steep, wall-like features called scarps. It appears that the craters on Iapetus may not have retained the full height of the crater rim, which would indicate the lithosphere is not thick or totally solid. The largest crater is 800 kilometers across with a rim topography of 10 kilometers (that is, raised 10 kilometers from the crater floor). Some scientists believe that Iapetus was formed and its shell hardened at a very early time. The surface might be the oldest surface known.

## Knowledge Gained

Iapetus has been studied with land-based telescopes since it was discovered. Its dichotomous albedo made it an

unusual object, and therefore an object of interest. Reflectance spectra taken using the McDonald Observatory produced data that showed reddening of the dark-side material, as dark-side reflectance was compared to that of the bright side. Other land-based studies determined the size of Iapetus, the albedo range, and the longitudinal symmetry of the dark material. One of the most in-depth studies of Iapetus's composition was done from the observatory at Hawaii's Mauna Kea, using both visible and near-infrared spectrometry.

The Voyager 1 mission revealed the diameter of Iapetus. Images of Iapetus were taken and revealed the great amount of cratering that exists on the bright side and at the poles. Perturbations in the path of Voyager allowed scientists to calculate the satellite's density.

In addition to the Image Science Substation (ISS), a version of which was on both Voyagers, the Cassini spacecraft (2004) had an ultraviolet imaging spectrograph (UVIS), which produces simultaneous spectral and spatial images. The visual and infrared mapping spectrometer (VIMS) was used to study the icy satellites by generating reflectance spectra and phase curves, as well as visual pictures. These data are especially important to compare with Earth-based data, providing an evaluation of the Earth-based data and possibly ideas of methods to correct the Earth-based data for two of the problems that occur using Earth-based instruments. Those two problems are the small phase angle seen from Earth and the extra light generated by reflection from the rings and from Saturn. The composite infrared spectrometer (CIRS) on Cassini studied the thermal infrared spectrum for emissivity features.

Certain compounds emit a thermal signature in the infrared range that is noticeable in the background thermal spectra. The thermal spectra from Iapetus did not show any strong features. The lack of emission features and the data from near-infrared spectra caused scientists to believe that the surface must be covered with small particles that have a high porosity. One bit of information gleaned from Cassini images is that scarps and crater walls are bright on their north sides and covered with dark material on their south sides.

## Context

Iapetus is certainly an enigma. The black face on one side and the bright face on the other side remain unexplained. Scientists have ideas for explaining these phenomena, but none that seems to offer a complete answer. If the dichotomy is due to material from other satellites, exactly what is the transport mechanism and what changes to the material occur during transport? A second question about Iapetus concerns its formation. How did the equatorial ridge form and why? What does the existence of the ridge tell scientists about the formation of the solar system? Is the surface of Iapetus the oldest surface in the solar system? Can Iapetus lead scientists to determine how the solar system was formed?

Many scientists are surprised that Iapetus can have such a large angle of inclination and be so far away from Saturn, yet still be in synchronous relationship with the planet. Being synchronous requires a very slow revolution, and such slow revolutions are unusual. Iapetus's revolution can be explained only if one concludes that the gravitational force of Saturn has caused despinning. The gravitational force on Iapetus increases and decreases as

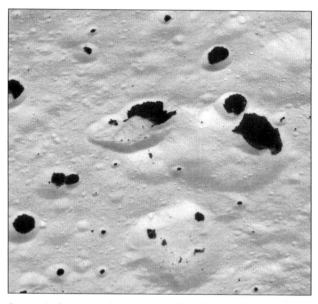

*Iapetus's frozen surface, splotched with dark terrain, in an image captured by the Cassini spacecraft on September 10, 2007, at an altitude of about 6,030 kilometers.* (NASA/JPL/Space Science Institute)

Iapetus gets closer and then moves away from the planet, and this change in gravitational force causes tidal flexing in the satellite. The heat from this tidal flexing may explain the heat needed to generate some of the satellite's features, such as the ridge and the dichotomous albedo.

      *C. Alton Hassell and David G. Fisher*

## Further Reading

Bond, Peter. *Distant Worlds: Milestones in Planetary* Exploration. New York: Copernicus Books, 2007. The author discusses several systems, including the Saturnian system and its parts: planet, satellites, and rings. Also discusses how various interplanetary missions have developed our knowledge of the systems from which they have gathered data. Illustrations, bibliography, appendix, index.

Castillo-Rogez, J. C., et al. "Iapetus' Geophysics: Rotation Rate, Shape, and Equatorial Ridge." *Icarus* 190, no. 1 (September, 2007): 179. Using the better data from Cassini, the authors describe several models for the formation of Iapetus. Each model is then evaluated against the known data. Illustrations, bibliography, index.

Corfield, Richard. *Lives of the Planets*. New York: Basic Books, 2007. The author takes the reader through the different space missions. Organized by planet. Index.

Fendrix, Amanda R., and Candice J. Hansen. "The Albedo Dichotomy of Iapetus Measured at UV Wavelengths." *Icarus* 193, no. 2 (February, 2008): 344. The dual face of Iapetus is discussed in view of measurements taken in the ultraviolet range. The meaning of a UV dichotomy is discussed. Illustrations, bibliography, index.

Hartmann, William K., and Ron Miller. *The Grand Tour: A Traveler's Guide to the Solar System*. 3d ed. New York: Workman, 2005. Each major planet, then the major moons including Iapetus, is discussed. Outstanding illustrations; bibliography, index.

McFadden, Lucy-Ann Adams, Paul Robest Weissman, and T. V. Johnson, eds. *Encyclopedia of the Solar System*. San Diego: Academic Press, 2007. The editors have collected articles written by many experts in one of the best scholarly reference works about the solar system. Illustrations, appendix, index.

Thomas, P. C., et al. "Shapes of the Saturnian Icy Satellites and Their Significance." *Icarus* 190, no. 2 (October, 2007): 573. Discusses how measurements from the Cassini mission reveal the shapes of the Saturnian satellites. Illustrations, bibliography, index.

# IO

**Categories:** Natural Planetary Satellites; The Jovian System

*Io is the innermost of four large satellites orbiting Jupiter, the largest planet of the solar system. Io is the most volcanically active body in the solar system. Tidal friction occurs constantly on Io, heating its core. Internal thermal energy is vented through immense volcanoes that spew sulfur and sulfur components into space which fall back, resurfacing the satellite. Io has one of the youngest surfaces in the solar system.*

## Overview

Until 1979, Io was known as little more than a pinpoint of light, even when seen through large telescopes. Io is one of the four satellites of Jupiter that were first observed telescopically by Galileo in 1609. With roughly the same size, density, and surface gravity as Earth's Moon, Io was expected to have similar features. Earth-based spectroscopic observations during the 1970's, however, raised speculation that Io was quite different. Suspicions were dramatically confirmed by the flybys of Voyagers 1 and 2 in 1979 and by multiple encounters of the Galileo spacecraft between December, 1995, and September, 2003. The closest encounter with Io, early in the Galileo mission, was at a distance of 22,000 kilometers; the probe approached no closer because of concerns about the intense radiation environment the spacecraft would encounter. During the late stages of the extended Galileo mission, however, the spacecraft passed within a mere 180 kilometers of Io's surface; at this late point in the mission, the science return was worth the risk to the spacecraft's health.

Voyager 1 made the remarkable, and completely unsuspected, discovery that active volcanoes dot the surface of this planet-like Jovian moon. From Voyager and Galileo evidence, augmented by Earth-based observations, scientists identified Io as a world dominated by volcanic eruptions. Most of the moon's surface features are transformed daily by heated liquid, with gaseous emissions onto the surface and into the otherwise nonexistent atmosphere. On Io, impact craters are not formed as they are on the other, extremely cold, satellites of the solar system. Io's impact features are apparently absorbed by molten lava flows that extend across Io's surface. Volcanic activity is not sporadic but virtually continuous. Of the eight active volcanoes observed by Voyager 1, seven were still spewing gaseous plumes when viewed by Voyager

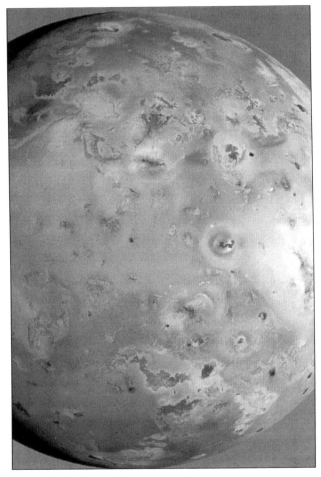

*The Galileo spacecraft captured this high-resolution image of Jupiter's moon Io from about 294,000 kilometers in March, 1998. (NASA/JPL/University of Arizona)*

2 four months later. Pictures taken by the two spacecraft portrayed these huge eruptions against the backdrop of the black sky over Io's limb (its visible horizon) and from above the red-orange surface with its active calderas. However, Io's volcanic activity changed in the time between the Voyager flybys and the arrival in orbit of the Galileo spacecraft on December 7, 1995.

With its image enhanced by spacecraft instruments, Io looked like a giant pizza, with wide plains of different hues punctuated by darker and lighter active regions. The latter are calderas, which dot the surface, at least two hundred of them, each having a diameter of more than 20 kilometers. The largest include eleven observed plume-emitting volcanoes named by the International Astronomical Union for mythological gods of early and primitive religions. Two of these, separated along the Loki fissure, are associated with a lava lake 200 kilometers wide, known as Loki Patera, which apparently is the major outlet for the planet's internal heat. Its temperature, like those of other Io "hot spots," averages about 300 kelvins. The hot spots contrast with the remaining 98 percent of the surface, which at 130 kelvins is considerably colder, as would be expected for an atmosphere-poor body so far from the Sun.

Mountains tend to cluster near the polar regions on Io. Some have peaks as high as 10 kilometers, but they do not appear to have been formed by plate tectonics (the shifting of continental geologic structures). They lack cone-like tops and could not have been formed by recent volcanism. It is speculated that some of Io's upper crust may detach and float about the molten plains in a manner analogous to icebergs. Also, erosion scarps form near emission calderas. The fluid surface suggests that Io might receive a new surface 10 micrometers thick every year, making it unique in a solar system of much older, inactive, heavily cratered planets and satellites.

The chemistry of Io helps explain this satellite's dynamic volcanism. Previously thought to have a solid interior, like other satellites of the solar system, Io appears to have a molten silicate core. Planetary geologists hypothesize that about four billion years ago, heated sulfur dioxide lying just below the surface became the driving force for the volcanoes, ejecting Io's internal heat in gaseous eruptions similar to geysers. Long-lived eruptions, like Loki, eject materials ballistically at 0.5 to 0.6 kilometer per second. Short-duration powerhouses, typified by Pele (and possibly Aten and Surt, seen by Voyager 2) do so at twice that velocity. These eruption rates are significantly higher than that of Earth's volcanoes (0.1 kilometer per second). Sulfur compounds are ejected as majestic mushroom-shaped plumes to a maximum height of 300 kilometers, enabling the lighter compounds, such as water and carbon dioxide, to escape into space. The heavier compounds, such as sulfur and sulfur dioxide, fall back to the surface as frozen, whitish, snowlike matter. Flows of molten matter therefore are low-viscosity sulfur and sulfur compounds rather than silicate rock lavas typical of Earth's volcanoes. Io's surface color results from the various sulfur compounds.

Io's geologically active behavior is caused by its proximity to massive Jupiter and to sister Galilean satellites Europa and Ganymede. Jupiter's gravitational pull causes Io to "flex" along its axis 10 kilometers toward Jupiter, while the combined attractions of Europa and Ganymede cause torques that give Io a slightly eccentric, noncircular orbit. The result is two opposing tidal forces stretching Io from within as it orbits Jupiter every 1.77 days. The ensuing friction raises Io's internal heated power to 60 to 80 trillion watts, partially melting the silicate compounds of the crust and generating volcanic eruptions. Furthermore, because Io's orbit lies entirely within Jupiter's radiation belts, the satellite is bombarded by charged particles and affected by the powerful electrical currents produced by Jupiter's magnetic field. These phenomena also influence Io's internal heating, as does the spontaneous radioactive decay of isotopes, which is typical of all planetlike bodies.

Io's volcanic emissions account for its irregular atmospheric pressure, first detected by Pioneer 10 in 1973. Atmospheric pressure variations result from the heat differential associated with the anomalous hot spots and the typically cooler surface. Earth-based observations in 1974 and 1975 detected a "cloud" of neutral sodium and potassium extending along Io's orbit for more than

*This amazing image of Io, taken by Voyager 1 in March, 1979, shows the eruption of a huge volcano at the upper left horizon—marking the first sighting of an active volcano beyond Earth.* (NASA)

100,000 kilometers. This observation was explained as a "sputtering" process whereby potassium atoms are ejected into space from Io because of the impact of charged particles from Jupiter's magnetosphere striking Io's surface. Their rate of ejection is greater than 10 kilometers per second, well above the necessary escape velocity of 2.5 kilometers per second. By comparison, Io's most active volcano, Pele, has an ejection speed of only 1 kilometer per second. Volcanoes, then, are an indirect rather than a direct cause of this sodium-potassium cloud; they bring the elements to the surface in molten and gaseous states for emission into space.

In 1976, additional Earth-based observations revealed a plasma "torus," or faint ring of excited glowing gas, belonging to Io's orbit. This torus occupies space within Jupiter's magnetosphere and results from the sputtering process. New electronic cameras and filters carried aboard the National Aeronautics and Space Administration's Kuiper Airborne Observatory recognized sulfur in 1981 and oxygen in 1982, as well as sodium and potassium, escaping from Io. The entire cloud assemblage supplies the torus with raw materials for further breakdown into discrete atoms and ionization. Thus energized, these ionized elements join the Jovian radiation belt that helped create them. The discovery of the torus was as unexpected as the volcanoes. Pioneer

10 and Voyager 1 flew directly through the torus in 1973 and 1979, respectively, but provided only knowledge supplementary to the major data obtained through continuous Earth-based monitoring. On its way to orbit insertion in December, 1995, the Galileo spacecraft flew a relatively safe distance from Io, one that had originally been considered to be the closest that Galileo would ever get to this innermost Jovian satellite, and then passed quickly through the plasma torus. No significant radiation damage was incurred by the spacecraft.

The volcanoes of Io offer the greatest promise for resolving the details of the complex relationship between Jupiter and Io. Of the eleven volcanoes discovered by Voyagers 1 and 2, the two-part vent of Loki is the most important. With a height of about 225 kilometers and a width of more than 430 kilometers, Loki and its lava lake, Loki Patera, appear to be the major outlet for Io's internal heat, as suggested by thermal emission polarization measurements in 1984. The thermal output from Io's greatest volcanoes, Pele (305 kilometers high and 1,200 kilometers wide), Surt, and Aten, is also significant. One of the three apparently ceased eruptions in 1986, according to Earth-based observations. The rest are 100 kilometers or less in height. Known calderas make up about 5 percent of Io's surface. Some, like Loki and Pele, have asymmetrical plumes and surface flows, which probably are consequences of irregular vent shapes. Others, like Prometheus, have symmetrical, fountainlike plumes and circular flows.

The Galileo spacecraft confirmed the massive scale of Io's volcanism, detecting a fresh volcanic deposit the size of Arizona and establishing that the satellite is rich in silicates. Increasingly sophisticated Earth-based instruments and orbiting telescopes have provided additional analysis. During eclipses in 1985, ice-covered Europa passed through Io's shadow, reflecting sunlight that revealed the uniform distribution of sodium about Io. The Hubble Space Telescope spotted an active plume in June, 1997, which Galileo also detected.

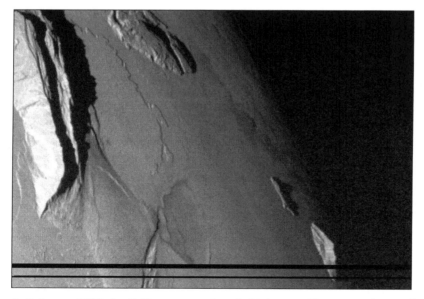

*In February, 2000, the Galileo spacecraft took this image of Io at a resolution of about 335 meters. To the left, Mongibello Mons rising about 7 kilometers from the surface. (NASA/JPL/University of Arizona/Arizona State University)*

### Knowledge Gained

Principally from Voyager 1 and Earth-based observations, Io was revealed to be the only volcanically active satellite

*An artist's conception of the role of Io's sulfur, which emerges from volcanoes and lands on the surface (see the arching arrow), where it is cooled into S3, S4 (both pinkish in color), and sulfur, S8, which gives the moon its characteristic yellow hue.* (NASA/JPL/Lowell Observatory)

of Jupiter. Volcanic gases and molten lava flows were seen being emitted from eleven major fissures. The most notable is the large dual vent of Loki Patera, which appears to be the focal point of the satellite's heat emissions from its interior. Because these eruptions are continuous, the ejected heavier compounds of material steadily move across the surface, eroding low-lying scarps and erasing the craters formed by objects striking Io. At least one lava lake was discovered, along with mountains near the poles. Emissions appear to be generated by Io's molten interior, from which sulfur compounds are emitted onto the surface and into the atmosphere. The heating mechanism for this activity is apparently internal friction caused by the gravitational pull of massive Jupiter on one side, and the satellites Europa and Ganymede on the other.

Because Io lies within Jupiter's magnetosphere, charged particles strike the satellite's volcanic surface, causing jets of potassium, sodium, sulfur, and oxygen to be ejected into its atmosphere—a "sputtering" process that feeds a cloud of those neutral elements. This cloud in turn supplies raw materials for a torus of excited gas along Io's orbital path.

Determination of Io's bulk density and moment of inertia has revealed the satellite to be a differential body composed of a silicate mantle with a metallic core that makes up as much as half of Io's radius. Comparative studies of the four Galilean satellites indicate that they share a number of similarities in their cores as well as differences due to their environment. Io, being active, has lost any significant icy shell it might have had over its core. Ganymede and Callisto, being colder and far less active, have thick, icy shells over their cores with craters on the icy surfaces of each. Europa has activity that gives it an icy crust that breaks up and flows over what is believed to be a liquid ocean beneath that crust.

As a result of Galileo's repeated observations of Io, hundreds of volcanic sites were identified, as were about two hundred dark surfaces believed to be fresh silicate lava. The orbiter's near-infrared mapping spectrometer and solid-state imager identified a hundred active hot spots through thermal emission.

## Context

Io emerged from the Voyager 1 and 2 missions as unique, not only among the satellites of Jupiter but also among all planets and satellites within the solar system. Io is recognized as the most volcanically active planetary body in a solar system where only a few other worlds display volcanism, and many of those display a type of cryovolcanic activity quite different in nature from Io's volcanoes. Io's volcanism is of particular interest because, like Earth, it is a dry body with a molten interior, and its sulfur-enriched chemistry may mimic volcanic conditions that existed during Earth's early history. Earth's active volcanoes convert water to steam for geothermal output. Io's sulfur-based volcanism provides an active laboratory for the study of planetary evolution, because volcanic eruptions constantly resurface its crust.

Io exists well within the Jovian magnetosphere. Electromagnetic fields affect Io's surface, allowing lighter elements emitted through its volcanic vents to escape into the atmosphere and feed the torus that encircles Jupiter as part of the radiation belt. The Jupiter-Io connection serves as a laboratory for the study of large-scale magnetic forces in the solar system. One of Io's effects on the Jovian system is a very gradual slowing of the rotation of Jupiter and erosion of the orbits of Europa and Ganymede.

Because active satellites like Io had not been anticipated before the 1979 Voyager flybys, their study has enhanced the evolving field of comparative planetology. Io was joined by Enceladus and Triton as satellites in the outer solar system that display unexpected volcanic activity.

Naturally, Io remains a high-priority target in planetary science for further study by robotic spacecraft. However, the radiation environment makes it difficult to dispatch probes into close proximity to the satellite. Human exploration is highly unlikely. There is a scene in Peter Hyams's film *2010: The Year We Make Contact* (1984) in which the wayward *Discovery* and its deactivated HAL 9000 computer are found adrift near Io. Spacewalking astronauts then transfer from a Russian spacecraft called the *Leonov* to enter and activate *Discovery*. In reality, the intense radiation environment would have provided such spacewalking astronauts a lethal dose well before they could return to the safety of another ship.

*Clark G. Reynolds*

## Further Reading

Bagenal, Fran, Timothy E. Dowling, and William B. McKinnon, eds. *Jupiter: The Planet, Satellites, and Magnetosphere*. Cambridge, England: Cambridge University Press, 2007. A comprehensive work about the biggest planet in the solar system. A series of articles provided by recognized experts in their fields of study. Excellent repository of photography, diagrams, and figures about the Jupiter system and the various interplanetary missions that have unveiled its secrets.

Beatty, J. Kelly, Carolyn Collins Petersen, and Andrew Chaikin, eds. *The New Solar System*. 4th ed. Cambridge, Mass.: Sky, 1999. A richly illustrated summary of early space-age discoveries that radically revised knowledge of the solar system, particularly useful in tracking initial reactions of scientists to those discoveries. Major features of Io and its volcanoes are covered.

Cole, Michael D. *Galileo Spacecraft: Mission to Jupiter*. New York: Enslow, 1999. Provides a full description of the Galileo spacecraft and science returns through the primary missions. Particularly good at describing mission objectives and goals. Suitable for a younger audience.

Fischer, Daniel. *Mission Jupiter: The Spectacular Journey of the Galileo Spacecraft*. New York: Copernicus Books, 2001. Thoroughly explains all aspects of the science and engineering of the Galileo spacecraft. Particularly good are the discussions about the nature of the Galilean satellites. Suitable for a wide range of audiences.

Geissler, Paul E. "Volcanic Activity on Io During the Galileo Era." *Annual Review of Earth and Planetary Sciences* 31 (May, 2003): 175-211. The definitive work describing the physics and planetary geology of volcanoes on Io. Provides a complete picture of Voyager and Galileo spacecraft results.

Greeley, Ronald. *Planetary Landscapes*. 2d ed. London: Allen and Unwin, 1994. A brief but instructive photographic examination of imaged planetary surfaces of the solar system, including a treatment of volcanism that sheds light on Io's unique volcanic processes.

Harland, David H. *Jupiter Odyssey: The Story of NASA's Galileo Mission*. New York: Springer Praxis, 2000.

This book provides virtually all of NASA's press releases and science updates during the first five years of the Galileo mission. Contains enormous numbers of diagrams, tables, lists, and photographs. Provides a preview of the Cassini mission. Published before the completion of the Galileo mission.

Hartmann, William K. *Moons and Planets*. 5th ed. Belmont, Calif.: Thomson Brooks/Cole, 2005. An updated version of a classic text that covers all aspects of planetary science. Takes a comparative planetology approach rather than including separate chapters on individual planets of the solar system.

Lopes, Rosaly M. C., and John R. Spencer. *Io After Galileo: A New View of Jupiter's Volcanic Moon*. Heidelberg: Springer, 2007. A volume in the Springer Praxis Space series, this book summarizes the knowledge gained by Galileo. Suggests new investigations needed to explain those questions that remain about the volcanism of this Jovian moon. Technical.

Morrison, David. "The Enigma Called Io." *Sky and Telescope* 69 (March, 1985): 198-205. An updated summary of original Voyager 1 and 2 data collected between 1979 and 1984, including contemporary information from Earth-based instruments, by a leading authority on the subject. Special attention is given to Loki Patera and the Jovian nebula, or thin gaseous torus, generated by Io.

Morrison, David, and Jane Samz. *Voyage to Jupiter*. NASA SP-439. Washington, D.C.: Government Printing Office, 1980. The official account of the Pioneer and Voyager flybys of the Jovian system, covering the day-to-day revelations from each mission. The most notable is the dramatic discovery of Io's volcanoes by Voyager 1. Lavishly illustrated.

# JOVIAN PLANETS

**Categories:** Planets and Planetology; The Jovian System

*Jupiter, Saturn, Uranus, and Neptune are called the Jovian planets. These "gas giants" have a mass 15-320 times greater than Earth, are of very low relative density, are mainly fluid (gas and liquid), and are composed of relatively light elements such as hydrogen and helium. All of them are surrounded by ring systems, a host of diverse satellites, and complex magnetospheres.*

**Overview**

The four Jovian, or "gas giant," planets are totally different geologically and physically from the terrestrial planets, Mercury, Venus, Earth, and Mars. Massive gaseous and liquid bodies composed primarily of hydrogen and helium, the Jovian planets are relatively rapid rotators. Each rotates about its own axis in less than twenty-four hours. The atmospheres of the Jovian planets—the feature that dominates observational work done on these planets—are very similar to one another in composition. Hydrogen represents about 90 percent of the atoms present, with helium making up the bulk of the remaining atmospheric gases. Methane and ammonia are also present, although ammonia on the two colder planets Uranus and Neptune has most likely precipitated out of the atmosphere. Weather systems that dominate these atmospheres, particularly in the case of Jupiter and Saturn, consist of rapidly rotating belts and zones that are visible from Earth. In the case of Jupiter, wind speeds on the order of 300 kilometers per hour are common, while on Saturn winds of two to three times that speed have been measured. Note that on Earth, hurricane-force winds rarely exceed 150 kilometers per hour. Ironically, wind speeds in the colder Uranus and Neptune are even higher than those seen on Jupiter and Saturn.

Both Jupiter and Saturn are much hotter than might be expected in view of their distances from the Sun. They have their own heat sources deep within their planetary interiors and thus are able to produce extensive thermal cells to drive high-speed winds. The nature of the heat within these planets is not entirely evident, and there is some evidence that even in the case of Uranus there may also be a heat-driven weather system resulting from a much more modest heat source on the planet. Uranus displays little atmospheric structure in visible light, but in ultraviolet there are some features. Images from Voyager 2 have shown that Neptune has an actively driven weather system. Dark storms and white streakers are seen to evolve in short time frames.

Although there are no measurements to indicate what lies below these turbulent atmospheres, there is indirect evidence that toward the center of a typical Jovian planet, pressures become higher. A portion of the interior is liquid. Toward the center of both Saturn and Jupiter, a very unusual state exists, that of liquid metallic hydrogen. This liquid metallic state would enhance both the thermal and electrical conductivity of the planetary interiors and no doubt is largely responsible for the strong magnetic fields associated with Jupiter and

Saturn. Pressures necessary to create liquid metallic hydrogen are on the order of millions of times the atmospheric pressure at the surface of Earth. Although they can be re-created in tiny cells in the laboratory, no such pressures have been sustainable in large-scale systems for prolonged periods of time on Earth. Thus, the liquid metallic hydrogen layer in the interiors of Jupiter and Saturn have effects that are not yet fully described. Uranus and Neptune probably have no such layers, for their masses are not great enough to produce such enormous interior pressures.

If one were proceeding inward toward a Jovian planet's center, one would next approach the core of that planet. Theorists disagree as to what might exist there, but the dominant opinion is that the cores would be largely solid and would contain relatively heavy metals such as iron, or they may be silicon in some high-pressure phase. In the case of Jupiter, such a solid core might have a mass 20 times greater than that of Earth, but this is a small fraction of the total mass of this planet, which is 320 times that of Earth. Uranus and Neptune might have solid cores on the order of several Earth masses, while that of Saturn would be about 5 to 10 Earth masses.

Little or nothing has been learned experimentally about planetary interiors; this is true even for Earth and its fellow terrestrial planets. Modeling planetary interiors, particularly on the scale necessary in the cases of Jupiter and Saturn, requires knowledge of pressure effects on bulk matter at pressures of millions of atmospheres and at alleged temperatures of 50,000 kelvins or hotter. It is known, however, that all the Jovian planets have a very low density, that is, a low specific gravity. Specific gravity is a measure of relative density, using water as a unit of 1 gram per cubic centimeter or 1,000 kilograms per cubic meter. Saturn has a specific gravity or relative density of 0.7; this means that it would float if one could find an ocean of water big enough in which to place it. It is by far the least dense of all the planets. Jupiter has an average density of 1.3 grams per cubic centimeter, while Uranus and Neptune have average densities of 1.2 and 1.7 grams per cubic centimeter, respectively. A typical terrestrial planet has a specific gravity of about 5. Earth's density is on average 5.5 grams per cubic centimeter. Overall, such relatively low density measurements indicate the predominance in Jovian planetary structures of light elements such as hydrogen and helium.

## Facts About the Jovian Planets

|  | *Jupiter* | *Saturn* | *Uranus* | *Neptune* |
|---|---|---|---|---|
| Mass ($10^{24}$ kg) | 1,898.6 | 568.46 | 86.832 | 102.43 |
| Volume ($10^{10}$ km³) | 143,128 | 82,713 | 6,833 | 6,254 |
| Equatorial radius (km) | 71,492 | 60,268 | 25,559 | 24,764 |
| Ellipticity (oblateness) | 0.06487 | 0.09796 | 0.02293 | 0.01708 |
| Mean density (kg/m³) | 1,326 | 687 | 1,270 | 1,638 |
| Surface gravity (m/s²) | 23.12 | 8.96 | 8.69 | 11.00 |
| Surface temperature (Celsius) | −140 | −160 | −180 | −200 |
| Satellites[a] | 63 | 60 | 27 | 13 |
| Mean distance from Sun millions of km (miles) | 779 (483) | 1,434 (891) | 2,872 (1,785) | 4,495 (2,793) |
| Rotational period (hrs)[b] | 9.9250 | 10.656 | −17.24 | 16.11 |
| Orbital period | 11.86 yrs | 29.66 yrs | 84.01 yrs | 164.79 yrs |

*Notes:*
a. Numbers are for known satellites as of the year 2009.
b. Retrograde rotational periods are preceded by a minus sign.
*Source:* Data are from the National Aeronautics and Space Administration/Goddard Space Flight Center, National Space Science Data Center.

Jupiter has a sizable magnetosphere. Its strong magnetic field is about ten times as intense as that of Earth. Jupiter's magnetosphere, which consists of trapped charged particles in amounts that would be lethal to humans, is so large that Saturn, which is 9.5 astronomical units (AU) from the Sun, passes through it. Saturn is almost twice as far from the Sun as is Jupiter (at about 5.2 AU), yet its magnetosphere is very strongly influenced by that of Jupiter. Saturn itself has a magnetic field slightly larger than that of Earth.

All the Jovian planets rotate about their own axes rapidly in relation to the terrestrial planets, which take at least twenty-four hours to make one rotation. (Earth is the fastest rotating terrestrial planet.) All Jovian planets thus exhibit some degree of oblateness. Saturn has an oblateness of about 0.1, which means that its equatorial diameter is about 10 percent bigger than its polar diameter, and thus it appears noticeably flattened at the poles. Saturn takes 10 hours and 13 minutes to make one complete rotation; Jupiter spins even faster, taking only 9 hours and 55 minutes to complete a rotation. Uranus takes 17 hours for one complete rotation, while Neptune takes about 16 hours (a figure that remains subject to conjecture in the aftermath of Voyager 2 studies in 1989).

These rapid rotations are surprising for such gaseous and liquid planets, because an angular momentum principle of elementary physics would have bigger bodies rotate more slowly than smaller ones. Such rapid rotation rates, then, are a mystery of the first magnitude in solar physics and geophysics. The correlation between magnetic fields and rotation rates is not very strong. Why some planets have powerful magnetic fields and others negligible ones is unknown. In general, however, if magnetic fields result from dynamo currents deep within planets, then rapid rotators should have strong magnetic fields. This is largely true in the case of both Jupiter and Saturn. Uranus has a magnetic field weaker than that of Saturn. Neptune has a magnetic field roughly comparable to that of Uranus.

Neptune and Uranus seem to be smaller and colder versions of Saturn and Jupiter. Uranus has a pale blue, almost greenish-blue appearance, undoubtedly because of the presence of methane. Imaging from Voyager photographs has shown that Uranus has belts and zones, although they are not as spectacular as those of Jupiter and Saturn. Ammonia within the Uranian atmosphere and that of Neptune is thought to have precipitated to the surface. In August, 1989, Voyager 2 flew by Neptune and returned images of that planet and its atmosphere. No good photographs of Neptunian surface features can be made from Earth's surface because of its distance (30 AU from the Sun). However, the Hubble Space Telescope has been used for intermittent studies of both Uranus and Neptune, some of the most productive research being performed by astronomer Heidi Hammel. Voyager 2 showed that Neptune is also a pale blue planet with irregular marked bands in its atmosphere. It has a gigantic Dark Spot, somewhat analogous to Jupiter's Great Red Spot, which seems to cause a tremendous sinking and upswelling of its atmospheric winds. This spot has a diameter about the same as the diameter of Earth. In addition, extremely high clouds, about 50 kilometers above the (normal) Neptunian atmosphere, make its atmosphere different from those of the other Jovian planets. Winds on the order of 200 meters per second have been measured in the Neptunian atmosphere, with unique streams and band systems.

All Jovian planets possess numerous moons, or satellites. By 2008, Neptune was discovered to have at least thirteen satellites. Uranus has at least twenty-seven detected moons, while Saturn has sixty and Jupiter has at least sixty-three. It is a bit surprising that these planets rotate so rapidly while at the same time each has such a large family of satellites. The presence of such satellites should have slowed the rotation rates, if indeed the satellites and the primary planets had common origins. Many of the satellites present the same face toward their primary planet, and thus are tidally locked.

All Jovian moons have rings as well, although they vary considerably in texture and content. The Voyager 1 spacecraft discovered a very thin ring around Jupiter in 1979. Since that time research has revealed that there is more structure to Jupiter's ring system. Presently it is referred to as the Halo Ring, the Main Ring, and the Gossamer Ring; the Gossamer Ring itself has two parts, the inner Almathea Gossamer Ring and the outer Thebe Gossamer Ring. Jupiter's rings are dark and composed of dust, and therefore they cannot be seen from Earth. Saturn's ring system of mainly icy particles is incredibly dynamic, with a number of gaps due to gravitational resonances and shepherding by small, embedded moonlets. Uranus has at least nine complete rings, some considerably brighter than others; it must be noted that five of Uranus's rings were discovered in the late 1970's by ground-based observations, not by the Voyager 2 mission, although the Voyager probe provided the first intense study of the entire ring system. It was Hubble-based research that found four additional dark rings long after the Voyager flyby. Neptune has only partial arcs. Astronomers are still debating whether partial arcs are ring systems being formed or in late stages.

## Methods of Study

Galileo's discovery in 1610 of the four large natural satellites of Jupiter (Io, Europa, Ganymede, and Callisto, known as the Medician and then later as the Galilean moons) launched the age of modern science. By the 1650's, Christiaan Huygens in Holland and other astronomers in Italy had established conclusively that Saturn had rings and at least one large satellite, Titan. The Great Red Spot on Jupiter has been observed ever since 1660, and the zones and belts on both Jupiter and Saturn had been clearly detected by enterprising visual astronomers. For three centuries, astronomers around the globe have tracked the Great Red Spot and noted changes in the belts and zones of these two gigantic planets. Uranus, too, was viewed by many from the 1700's onward, but it was not clearly designated as a planet until the late eighteenth century.

With the advent of spectroscopy in the nineteenth century, helium was discovered first on the Sun and shortly thereafter on Jupiter and Saturn. In the early twentieth century, it was learned that methane and ammonia were present in the Jovian atmospheres as well. In 1955, radio astronomers detected radio signals coming from Jupiter's magnetosphere.

It was not until the 1970's, however, that the greatest discoveries about the Jovian planets were made. Data from Voyager 1 revealed the unexpected existence of rings around Jupiter. Earth-based observations showed rings around Uranus and Neptune, and images returned from Voyager 2 in the 1980's revealed ten satellites circling Uranus and orbiting Neptune.

Radio astronomy probes the decimetric and decameter radio signals emitted from Jupiter, which signal the extent of its magnetosphere and the relationship of its halo and its volcanic innermost satellite Io, respectively. Radio astronomy conducted by the Cassini spacecraft in orbit about Saturn provided insight into the nature of the magnetic field on Enceladus. Infrared astronomy also has been very helpful in determining some of the features of the cold Jovian planets Uranus and Neptune. Many experimental techniques have been used to determine the size and extent of the ring systems surrounding these planets, and still there are many unanswered questions about these systems.

The atmospheres of Jupiter and Saturn have been probed with all sorts of sensitive spectrometers, yet experimental information is valid only for a penetration depth of a few tens of kilometers. What lies below the turbulent, fast-moving atmosphere has not been experimentally detected; all that scientists can do is rely on the best theories and modeling techniques presently available.

### Jovian Planets' Atmospheres: Comparative Data

|  | *Jupiter* | *Saturn* | *Uranus* | *Neptune* |
|---|---|---|---|---|
| Surface pressure (bars) | >100 | >100 | >100 | >100 |
| Surface density (kg/m3) | ~0.16 | ~0.19 | ~0.42 | ~0.45 |
| Avg. temperature (kelvins) | ~129 | ~97 | ~58 | ~58 |
| Scale height (km) | 27 | 59.5 | 27.7 | ~20 |
| Composition |  |  |  |  |
| Ammonia | 260 ppm | 125 ppm | tr | tr |
| Ethane | 5.8 ppm | 7 ppm | — | 1.5 ppm |
| Helium (%) | 10.2 | 3.25 | 15.2 | 19 |
| Hydrogen (%) | 89.8 | 96.3 | 82.5 | 80 |
| Hydrogen deuteride (ppm) | 28 | 110 | ~148 | 192 |
| Methane | 3000 ppm | 4500 ppm | ~2.3% | 1.5% |
| Water | — | ~4 ppm | tr | tr |

*Notes:* Composition: % = percentages; ppm = parts per million; tr = trace amounts.
*Source:* Data are from the National Space Science Data Center, NASA/Goddard Space Flight Center.

The Pioneer 10 and 11 probes found that Jupiter is a tremendous source of electrons and that it generates several times as much heat as it receives from the Sun. The origin of these electrons and heat is far from clear to the most discriminating theorists in both physics and geophysics. There are no comparable conditions on Earth or the nearby terrestrial planets to produce such effects. Voyager 2 passed Neptune in August, 1989, and its use as a planetary probe effectively ceased. Its next primary objective was to characterize the approach to interstellar space. The Hubble Space Telescope (HST) has produced much better images of Uranus and Neptune than had previously been available on a regular basis.

The Galileo spacecraft arrived at Jupiter in 1995. A special probe released from Galileo entered Jupiter's atmosphere on December 7, 1995, and its instruments detected a new radiation belt, fierce winds, lightning, and upper-atmosphere densities and temperatures much higher than expected. The Cassini spacecraft, launched in October, 1997, began exploring the Saturn system from an orbital vantage point beginning in 2004. The Huygens probe that Cassini carried along on its journey from Earth to Saturn was released and sent down through the atmosphere of Saturn's largest satellite, Titan, a moon with a thick atmosphere that obscures its surface. Cassini carried an imaging radar to map the satellite's surface during repeated close flybys. Huygens survived its plunge through the atmosphere and landed in a mushy, cryogenic surface. Huygens sent its data on two redundant channels, but, because of a software error, only one transmitted properly; fortunately, an alternative path recovered most of the data that otherwise could have been lost. Huygens and Cassini found evidence of complex hydrocarbons under cryogenic conditions on the surface of Titan. Cassini's primary mission was completed in 2008, and the program received a fully funded two-year extension.

## Context

It was expected that early exploration of the Jovian planets and their extensive satellite systems would provide scientific clues as to how the solar system formed. Instead, a whole series of new mysteries has appeared, spurring additional robotic exploration of the outer solar system.

For atmospheric physicists, the weather systems evident in the atmospheres of both Jupiter and Saturn have provided much material for study. The Great Red Spot and several of the lesser white spots on both Saturn and Jupiter have proved to be cyclonic or anticyclonic storms that are able to maintain themselves for decades. The Great Red Spot has been observed for at least for 350 years. Could such a massive storm system be maintained on Earth? What conditions on Jupiter contribute to the tremendous longevity of the Great Red Spot? In attempting to answer questions of this sort, scientists have modeled all manner of weather systems, which has proved to be useful in deciphering meteorological patterns on Earth. Thus Jupiter and Saturn have served as gigantic, high-pressure, turbulent laboratories for atmospheric modelers. Indeed, the greatest potential outcome of comparative planetology is a better understanding of complex geophysical and atmospheric physics processes right here on Earth. Such an understanding is fundamental to determining whether or not Earth is presently undergoing global warming of natural or human-made origin.

Even in the esoteric discipline of fluid mechanics, particularly in the study of turbulent flow, data from Jupiter and Saturn have been unexpectedly helpful. These studies are critical in air-frame design and, when coupled with modern computer modeling techniques, have proved to be very valuable in the design of supersonic air frames and high-speed hydrofoils. Neptune's Dark Spot should provide fodder for studies of both fluid-mechanics and meteorology well into the twenty-first century. Many scientists believe that solar-system locations most likely to host life or organic chemistry necessary for life are either Jupiter's ice-covered satellite Europa, or Saturn's satellites Titan and Enceladus. Some sort of life systems could be operating in either of these locations, for the energy and chemical conditions seem suitable. Should some sort of complex organic molecules or anaerobic bacteria be found on either Jupiter or Europa, the perennial mystery as to how life formed on Earth and why it exists at all could be addressed intelligently, perhaps for the first time. Saturn's satellite Enceladus has cryogenic geysers in its south polar region. Neptune's satellite Triton has also appeared to contain some sort of cryogenic geyser activity. The unexpected detection of warm liquids in the outer solar system could drive biological networks. Titan has organic materials that are believed to be indicative of the primordial Earth, although the satellite is far colder than Earth was when life developed here. Thus Titan might be a frozen example of what the early Earth might have been when life first arose.

In 1955 radio astronomers Bernard Burke and Kenneth Franklin, while studying the Crab nebula, inadvertently discovered radio emissions coming from Jupiter. Decades later, Voyager 2 recorded the largest electrical current ever measured as it passed near Jupiter. In the first half of the twentieth century, most scientists did not realize that Earth, with its reasonably strong magnetic field, produced a magnetosphere just as Jupiter did. It was not until early American spacecraft discovered the Van Allen radiation belts that radio engineers, astronomers, and plasma physicists realized that Earth's magnetosphere was a smaller version of Jupiter's. The magnetospheres of Jupiter, Saturn, and even Earth are still not completely understood; what influence they might have had on planetary origins and developments is unknown. Eventually, studies of Jupiter and Saturn might provide clues regarding the forces and mechanisms behind electrical storms, violent atmospheric electricity, and radio blackouts that can have pronounced effects on life on Earth.

*John P. Kenny*

## Further Reading

Bagenal, Fran, Timothy E. Dowling, and William B. McKinnon, eds. *Jupiter: The Planet, Satellites, and Magnetosphere*. Cambridge, England: Cambridge University Press, 2007. A comprehensive work about the biggest planet in the solar system, including a series of articles provided by recognized experts in their fields of study. Excellent repository of photography, diagrams, and figures about the Jupiter system and the various spacecraft missions that unveiled its secrets.

Bortolotti, Dan. *Exploring Saturn*. New York: Firefly Books, 2003. A look at the Cassini-Huygens mission for a younger audience. Full of charts, photographs, a section on observing Saturn, and an overview of the history of our understanding of the Saturnian system, from antiquity to the launch of Cassini.

Greenberg, Richard. *Europa the Ocean Moon: Search for an Alien Biosphere*. New York: Springer, 2005. A complete description of Europa through the post-Galileo spacecraft era. Discusses the astrobiological implications of an ocean underneath Europa's icy crust. Well-illustrated and readable by both astronomy enthusiasts and college students.

Harland, David H. *Jupiter Odyssey: The Story of NASA's Galileo Mission*. New York: Springer Praxis, 2000. Collects virtually all of NASA's press releases and science updates during the first five years of the Galileo mission, along with a preview of the Cassini mission. Includes an enormous number of diagrams, tables, lists, and photographs.

Harland, David M. *Cassini at Saturn: Huygens Results*. New York: Springer, 2007. Provides a thorough explanation of the entire Cassini program, including the Huy-gens landing on Saturn's largest satellite. Essentially a complete collection of NASA releases from the start of Cassini flight operations through the majority of Cassini's seventy orbits during its primary mission (Cassini's primary mission concluded a year after this book was published). Technical writing style but accessible to a wide audience.

_____. *Mission to Saturn: Cassini and the Huygens Probe*. New York: Springer Praxis, 2002. A volume in Springer's Space Exploration series, this is a technical description of the Cassini program, its science goals, and the instruments used to accomplish those goals. Written before Cassini arrived at Saturn. Provides a historical review of pre-Cassini knowledge of the Saturn system.

Hartmann, William K., ed. *Astronomy*. 5th ed. Belmont, Calif.: Wadsworth, 2004. Hartmann's section of this astronomy textbook, which should be accessible to high school and college students, examines the Jovian planets and other parts of the solar system. Besides discussing many late twentieth century findings, Hartmann lists various theories of planetary origins and natures, and he examines the strengths and weaknesses of each. One chapter focuses on Jupiter, and another compares Jupiter to the other Jovian planets. Well illustrated.

Irwin, Patrick G. J. *Giant Planets of Our Solar System: An Introduction*. 2d ed. New York: Springer, 2006. Suitable as a textbook for upper-level college courses in planetary science. Focuses on Jupiter, Saturn, Uranus, and Neptune and their satellites, rings, and magnetic fields. Filled with figures and photographs.

Karttunen, H. P., et al., eds. *Fundamental Astronomy*. 5th ed. New York: Springer, 2007. A well-used university textbook in introductory astronomy. Contains some calculus-based treatments for those who find the standard texts for introductory astronomy too low-level. Covers all topics from solar-system objects to cosmology.

Lovett, Laura, Joan Harvath, and Jeff Cuzzi. *Saturn: A New View*. New York: Harry N. Abrams, 2006. A coffee-table book with about 150 of the best images returned by the Cassini mission to Saturn. Covers the planet, its many satellites, and the complex ring system.

Russell, Christopher T. *The Cassini-Huygens Mission: Orbiter Remote Sensing Investigations*. New York: Springer, 2006. Provides a thorough explanation of the remote-sensing investigations of both the Cassini orbiter and the Huygens lander. Outlines the scientific objectives of all instruments on the spacecraft and describes the planned forty-four encounters with Titan.

Given the publication date, only early science returns are discussed.

# JUPITER'S ATMOSPHERE

**Categories:** Planets and Planetology; The Jovian System

*Jupiter's atmosphere differs greatly from that on Earth. It is composed mainly of hydrogen and helium and is far enough from the Sun that the temperature of the visible cloud deck is only 153 kelvins. Voyager spacecraft data revealed details concerning chemical composition, heat transport, and wind patterns within the atmosphere. The Galileo spacecraft's atmospheric entry probe sampled that atmosphere directly and forced a rethinking of the physical model of Jupiter.*

## Overview

Observed through an Earth-based telescope, Jupiter's most striking aspects are a pearly glow reflected from the planet; a series of east-west bands of pastel yellows, whites, browns, and blues; and the oblate aspect of its disk. The equatorial diameter is 6 percent larger than the polar diameter, a fact that is readily apparent to the observer. Closer inspection will reveal that distinctly bright and darker individual cloud features are visible within the banded structure of the atmosphere. Throughout an evening, an observer will notice that cloud features are rotating from west to east at a rate of about 36° per hour, indicating that Jupiter rotates on its axis in slightly less than ten hours. A careful student of Jupiter will discover that the planet's visible cloud deck does not rotate as a solid body; instead, eastward winds at the equator sweep the clouds past those at midlatitudes at a rate that displaces them 7° eastward each day. This type of motion indicates that the cloud deck is opaque. In order to understand the atmospheric winds, one must know how fast the core of the planet is rotating.

Radio astronomers collected data from Jupiter and realized that variations in the radio signals from Jupiter could be attributed to the interaction of the charged particles ejected from the Sun with the magnetic field of the planet. Although the signal that the radio astronomers were measuring was generated above the atmosphere of the planet, the signal varied as Jupiter's magnetic field rotated. Astronomers' understanding of magnetic fields led them to believe that the radio astronomers were measuring the rotation of Jupiter's core. They determined that one rotation took 9 hours, 55 minutes, and 29.771 seconds.

A ground-based observer equipped with an eyepiece outfitted with a crosshair can center the planet and record the time that a selected cloud feature takes to rotate past the crosshair. Because the planet rotates about its axis in approximately ten hours, while the observer is constrained to observe within a twenty-four-hour time frame, the feature will be visible on alternate nights. The rate of rotation of the feature across the visible disk of the planet,

*The Cassini spacecraft delivered four images in December, 2000, that were compiled to produce this picture of Jupiter with a resolution of 144 kilometers.* (NASA/JPL/ University of Arizona)

however, is such that the observer will find that the cloud rotates five times in slightly more than two days.

Accurate periods of rotation were determined by the British Astronomical Association and the American Lunar and Planetary Institute during the first half of the twentieth century. Careful measurements of photographic data by Elmer Reese from 1960 to 1974 refined these data. When the data were related to the planetary core using the radio period of rotation, an alternating pattern emerged. Strong eastward winds near the equator, as swift as 150 meters per second (more than 450 kilometers per hour), decreased poleward. Near 15° latitude, the prevailing wind in both hemispheres was westward. Between 10° and 35° latitude, the displacement of clouds revealed two westward and two eastward peaks in the horizontal wind speeds. Highly reflective regions called zones were bracketed on the equatorward side by westward winds and on the poleward side by eastward winds. The less reflective, browner regions, or belts, were nested between the zones. Horizontal wind flow of this type in Earth's atmosphere would generate conditions that would cause rising air in the zones, leading to the formation of ice clouds at high altitudes. Air would descend in the belts and cause ices to melt, allowing a longer line-of-sight through the atmosphere and more absorption of light—hence, less reflection. Recognition of this general circulation pattern in the 1960's led to questions concerning the nature of Jupiter's ices.

As light travels outward from the Sun, it spreads out equally; thus, its ability to heat a surface decreases rapidly via an inverse square relationship in all radial directions. By the time sunlight reaches Jupiter, at a distance five times greater than Earth's distance from the Sun, the intensity is diluted by a factor of twenty-five. This dilution leads to temperatures too low to allow melting of ice formed from water; therefore, the visible cloud deck must contain another kind of ice.

In an effort to refine their understanding of the temperature regime, astronomers began to make infrared measurements of Jupiter's atmosphere. They determined that the planet radiates one and a half times more heat than it absorbs from the Sun. These results imply that the interior of Jupiter is hotter than the cloud deck and that convection from the interior transports heat outward. The picture of a deep atmosphere dominated by east-west winds emerged. If Jupiter were composed of the same chemical mixture as the Sun, its atmospheric gas would be so strongly compressed that deep below the visible cloud deck it would form a sea of liquid hydrogen and helium. There would

be an indistinct change between the surface of the cryogenic fluids and the atmosphere. It also became apparent that it was essential to understand the chemistry of the atmosphere.

Calculations carried out by John Lewis and Ronald Prinn led to a model of the atmosphere that posited the existence of an upper cloud layer of ammonia ice, underlain by an ammonium hydrosulfide layer and a cloud layer composed of water ice (where the pressure is greater than ten times Earth's atmospheric pressure at sea level). All these ices are white; therefore the calculations yielded no information about coloring agents in the Jovian atmosphere. Above the topmost cloud deck, the atmosphere is mainly hydrogen and helium gas, with traces of ammonia and methane.

Molecules of methane and ammonia can absorb enough energy from incident ultraviolet light to break bonds, which allows the hydrogen atoms to escape. Darrel Strobel performed calculations indicating that ionized molecules combine to form more complex molecules and possibly aerosols or hazes. Laboratory work by other investigators has shown that many of the compounds that are formed are yellow and brown. Much of the color variation in the Jovian atmosphere may arise from variations in the heights of the underlying clouds.

Removal of the smog occurs in two ways. Either the particles grow large and fall to lower levels, or convective clouds of ammonia are carried upward like thunderheads, and the ammonia ice encases the smog particles and again causes them to fall to lower levels in the atmosphere. Computer modeling of scattering and transmission of light through the layers of hazes, gas, and clouds, carried out by Martin Tomasko, Robert West, and others, supports the theory that colorization is dependent on the varying heights of underlying clouds.

These calculations cannot explain Jupiter's Great Red Spot—neither its colors nor its long life. This feature is the largest Jovian cloud system. North to south, it spans a distance slightly larger than the diameter of Earth, and it extends farther than two Earth diameters in the east-west direction. This large, unique cloud system is trapped between a westward wind on the equatorward side and an eastward wind on the poleward side. The winds are diverted around its perimeter, rotating it in a counterclockwise direction. In Earth's atmosphere, a weather system with similar motion would rise in the center, spiral outward at the top of the cloud deck, and descend around its perimeter. The degree of redness of the Spot varies with time. A unique property of the Red Spot is its ability to

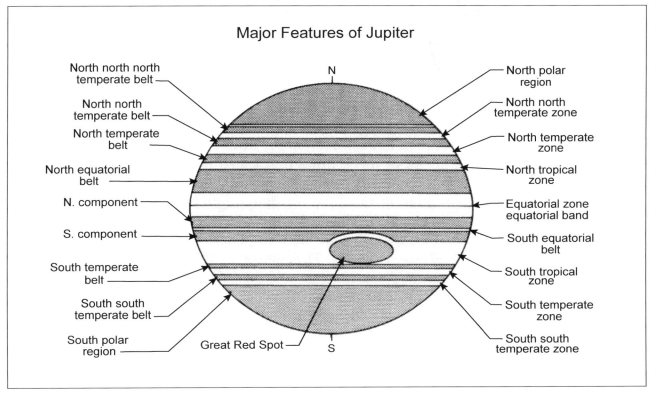

Major Features of Jupiter

*Source:* Morrison, David, and Jane Samz. Voyage to Jupiter. NASA SP-439. Washington, D.C.: National Aeronautics and Space Administration, 1980, p. 4.

absorb ultraviolet, violet, and blue light. Apparently some constituent that has been carried up from lower, warmer depths absorbs the blue light, causing the Spot to appear redder than the surrounding clouds.

Observations of Jupiter in the infrared indicate that brown and blue-gray regions are warmer than white areas. Thus, the white zones are colder than the brown belts. The clouds above the Red Spot are cold, but there is a warmer region around its perimeter. The general heat loss from the planet indicates that the interior is warmer than the upper cloud deck that radiates to space; hence, infrared maps allow astronomers to determine relative heights of clouds. A desire to obtain high-resolution maps of the infrared data and the structure of the cloud deck led to the development of the instruments that were on board the Pioneer and Voyager missions. Perhaps the best way to determine a planet's atmospheric character is to directly encounter it. That was the reasoning behind having the Galileo spacecraft carry an atmospheric probe that then was dispatched to

hit the upper atmosphere and relay data until tremendous pressures destroyed it.

### Knowledge Gained

The Pioneer 10 and 11 spacecraft, which arrived at Jupiter in November, 1973, and November, 1974, respectively, each carried three instruments that sampled the Jovian atmosphere. The fact that the Pioneer spacecraft were spin-stabilized limited the types of instruments that could be placed on board. Voyagers 1 and 2 passed through the Jupiter system four months apart in 1979. Ultraviolet and infrared instruments and a spin-scan camera were trained on Jupiter's atmosphere. The Galileo probe also sampled the atmosphere beginning in 1995. The Cassini spacecraft in December, 2000, and the New Horizons spacecraft in February, 2007, conducted in-depth studies of Jupiter as they flew past on their way to Saturn and Pluto, respectively.

Infrared data revealed that there was little temperature variation between the equator and the pole at the cloud-top level. Data indicated there are limits to the role that equatorial solar heating plays in driving the zonal winds. Andrew Ingersoll proposed that solar heating at the equator could bring about a cloud structure that would act as an insulating blanket, causing the heat from the interior to emerge near the poles and resulting in little temperature variation at the level of the ammonia cloud deck. This hypothesis implies that the outward transport of the heat from the interior may dominate the atmospheric wind patterns.

Pioneer spin-scan cameras were equipped with blue and red filters and polarizers. The nature of the camera did not allow a large number of images to be obtained. Nevertheless, data provided valuable material for study of the scattering properties of the atmospheric smog and haze layers. Although a series of images with sufficiently high resolution to map cloud motions could not be obtained, images of the Red Spot and north polar regions confirmed information previously gained from ground-based observation. They also provided data on the scale of the cloud structures.

Voyagers 1 and 2 carried five instruments that were used to observe Jupiter's atmosphere: two television cameras (one with a wide-angle view and the other with higher resolution and a narrow field of view), infrared and ultraviolet spectrographs, and a photopolarimeter. Multicolor high-resolution mapping of the visible cloud deck could be obtained at three-month intervals with each spacecraft. Near-encounter infrared measurements resolved temperature variations as a function of latitude and longitude on the planet. The ultraviolet spectrometer obtained data concerning high altitudes in the Jovian atmosphere. This extensive data set has been combined with the Pioneer and historical ground-based data sets in an effort to shed light on both short-term and long-term atmospheric variations. Cloud displacements were measured by Reta Beebe, Ingersoll, and others. Eastward winds near the equator were as powerful as 160 meters per second. Westward wind speeds at 15° north and 17.5° south latitude were both retrograde at 40 and 70 meters per second, respectively. Eastward wind

maxima at 20° north and 24.5° south latitude were 170 and 60 meters per second, respectively. Voyager scientists found considerable differences between the magnitudes of wind jets in the northern hemisphere and those in the southern hemisphere; no change in the average zonal wind was detected at any latitude, however, during the five-month interval between the two encounters.

Infrared measurements indicated that the winds decrease with height above the deck, and that temperatures and abundance of ammonia above the cloud deck are consistent with an atmosphere that is driven by cloud motions at the level of the visible cloud deck.

Galileo's entry probe hit Jupiter's upper atmosphere on December 7, 1995, at a speed of as much as 170,000 kilometers per hour. The atmosphere decelerated the probe at an increased g-load of approximately 230 (that means the force was 230 times that of normal Earth gravity at sea level). During the probe's 57-minute-long plunge, it successfully relayed its findings to the Galileo orbiter for storage and eventual playback to Earth. Galileo was a little over 200,000 kilometers above the probe at the time.

The probe provided some surprising data. It had been hoped that the probe would find considerable

*Voyager 1 captured this image of Jupiter's roiling atmosphere in 1979.* (NASA/JPL)

amounts of water vapor in the atmosphere and detect extensive amounts of electrical activity, or lighting. In reality it found very little of either. It did find a new radiation belt just 50,000 kilometers above the cloud tops. As it descended through Jupiter's atmosphere, the probe registered very strong winds and experienced significant turbulence. Spectrometers found lower abundances of helium, neon, carbon, oxygen, and sulfur than had been expected. Helium was nearly half as abundant as expected in contemporary atmospheric models for Jupiter; Galileo researchers were expecting the probe to fall through a three-layered cloud structure. The probe did not experience anything like what had been predicted. The net flux radiometer on the probe did find some high-level ammonia ice clouds, and the nephelometer instrument provided some evidence of ammonium hydrosulfide clouds. Water ice was absent, suggesting the probe had entered one of the driest spots on Jupiter.

Wind strengths and atmospheric temperatures varied during the probe's descent. Winds reached 350 kilometers per hour with gusts up to 525 kilometers per hour. After the probe had plunged 156 kilometers through the atmosphere under its main parachute, the temperature and pressure environment destroyed it; most likely it was crushed, vaporized, or both nearly simultaneously. Essentially the probe's encounter forced planetary scientists to rethink the current model of Jupiter's atmosphere.

## Context

By the mid-nineteenth century, astronomers had become aware that Jupiter was unlike Earth. Using the basic laws of motion, the apparent size of Jupiter, and the known distances within the solar system, they determined that even though the volume of Jupiter was more than eleven hundred times larger than Earth's volume, its mass was only 318 times larger than that of Earth. Thus, although this planet is much more massive than Earth, gravity has not compressed Jupiter's interior to the high densities that are present in the interior of Earth. Nineteenth century astronomers concluded that Jupiter could not have the same chemical composition as Earth.

By 1960, spectroscopists had determined that the atmosphere of Jupiter was cold and that temperatures at the level of the visible clouds were near 153 kelvins. Spectra revealed absorption by molecules of methane and ammonia. These observations are consistent with an atmosphere that is composed mainly of hydrogen and helium with small amounts of carbon and nitrogen. At the observed temperatures, oxygen would combine with hydrogen to form water, which would be trapped below the visible cloud deck. It became apparent that Jupiter was composed of a chemical mixture similar to that of the Sun, and that the small silicon- and iron-rich planets of the inner solar system were very different from the outer gas-rich planets.

Current models concerning the formation of a solar system propose that planets the size of Jupiter form first at distances far enough from the parent star that hydrogen and helium have not been expelled by the radiation from the star. The turbulence that this generates in the preplanetary gas and dust cloud leads to the formation of the other planets, with the inner ones forming from hydrogen-poor material. The importance of Jupiter-sized bodies in the formation of other planets that could possibly support other life-forms has stimulated interest in learning more about the nature of this gas giant. However, it must be noted that extrasolar planets have been found to have masses in excess of Jupiter and to be located extremely close to their stars.

Jupiter's atmosphere is chemically unlike that of Earth. The planet's huge depth and lack of irregular landmasses at its lower boundary contrast with conditions in Earth's atmosphere. There are, however, some similarities: The main constituents of Jupiter's atmosphere are hydrogen and helium. Like the nitrogen and oxygen molecules of Earth's atmosphere, these particles do not absorb sunlight readily. A large portion of solar energy passes through the upper atmospheres of these planets and, in the case of Earth, the surface absorbs the energy and is warmed. The atmosphere is then heated from the bottom, with trace constituents, carbon dioxide, and water absorbing and reradiating the energy. This leads to decreasing temperatures at increasing altitudes in the lower atmosphere.

Voyager infrared data confirmed that Jupiter has an internal heat source and that it emits 1.67 times more energy than it absorbs from the Sun. Infrared data indicate that the winds of Jupiter are driven, like those on Earth, by energy input in the lower atmosphere. Knowledge of planetary atmospheres has become more general and efforts to define the factors that lead to climate variations can be applied to more than one planet. Jupiter's dissimilarity to Earth provides checks and challenges in the search to understand Earth and the solar system.

*Reta Beebe*

## Further Reading

Bagenal, Fran, Timothy E. Dowling, and William B. McKinnon, eds. *Jupiter: The Planet, Satellites, and Magnetosphere.* Cambridge, England: Cambridge University Press, 2007. A comprehensive work about the biggest planet in the solar system, covered in a series of articles provided by recognized experts in their fields of study. Excellent repository of photography, diagrams, and figures about the Jupiter system and the various spacecraft missions that unveiled its secrets.

Beatty, J. Kelly, Carolyn Collins Petersen, and Andrew Chaikin, eds. *The New Solar System.* 4th ed. Cambridge, Mass.: Sky, 1999. Filled with color diagrams and photographs, this popular work covers solar-system astronomy and planetary exploration through the Galileo missions. Accessible to the astronomy enthusiast.

Cole, Michael D. *Galileo Spacecraft: Mission to Jupiter.* New York: Enslow, 1999. Provides a full description of the Galileo spacecraft, its mission objectives, and science returns through the primary mission. Particularly good at describing mission objectives and goals. Suitable for a younger audience.

Fimmel, Richard O., James Van Allen, and Eric Burgess. *Pioneer: First to Jupiter, Saturn, and Beyond.* NASA SP-446. Washington, D.C.: Government Printing Office, 1980. A detailed review of the original Pioneer mission. Reproduces most of the images obtained by the spin-scan camera. Suitable for the general reader.

Gehrels, Tom, ed. *Jupiter.* Tucson: University of Arizona Press, 1976. A historic collection of scientific essays covering all aspects of Jupiter. The volume reflects the state of knowledge of the planet before the Voyager mission; a subsequent survey has not been published. Its ample documentation will, however, direct the serious student to journals and other sources that update the information available here. For the advanced reader.

Harland, David H. *Jupiter Odyssey: The Story of NASA's Galileo Mission.* New York: Springer Praxis, 2000. This book provides virtually all of NASA's press releases and science updates during the first five years of the Galileo mission. Provides a preview of the Cassini mission. Includes an enormous number of diagrams, tables, lists, and photographs. The book's description ends before completion of the Galileo mission unfortunately, but what is missing can easily be found on numerous NASA Web sites.

Hartmann, William K. *Moons and Planets.* 5th ed. Belmont, Calif.: Thomson Brooks/Cole, 2005. An updated version of a classic text that covers all aspects of planetary science. The chapter on Jupiter thoroughly addresses the Jovian system and spacecraft exploration of it.

Hunt, Garry E., and Patrick Moore. *Jupiter.* New York: Rand McNally, 1981. Reviews the original Voyager mission and describes the Jovian system. Photographs and illustrations are plentiful. Requires some background knowledge.

Irwin, Patrick G. J. *Giant Planets of Our Solar System: An Introduction.* 2d ed. New York: Springer, 2006. Focuses on Jupiter, Saturn, Uranus, and Neptune and their satellites, rings, and magnetic fields. Filled with figures and photographs. Suitable as a textbook for upper-level college courses in planetary science.

McAnally, John W. *Jupiter, and How to Observe It.* New York: Springer, 2008. An observing guide for the amateur astronomer that also provides detailed descriptions of the Jovian system. Discusses observational techniques, including a wide range of popular telescopes and ancillary equipment.

McBride, Neil, and Iain Gilmour, eds. *An Introduction to the Solar System.* Cambridge, England: Cambridge University Press, 2004. A complete description of solar-system astronomy suitable for an introductory college course. Accessible to nonspecialists as well. Filled with supplemental learning aids and solved student exercises. A companion Web site is available for educator support.

Morrison, David, and Jane Samz. *Voyager to Jupiter.* NASA SP-439. Washington, D.C.: Government Printing Office, 1980. A description of the events surrounding the Voyager missions to Jupiter. Illustrated with many color photographs. Accessible to the general reader.

Peek, Bertrand M. *The Planet Jupiter.* London: Macmillan, 1958. A detailed summary of ground-based observations recorded by the British Astronomical Association. This classic book provides an overview of the time-dependent aspects of the Jovian cloud deck and the history of the Red Spot. Although dated, this accessible presentation of the basics can be compared with more contemporary understandings.

# JUPITER'S GREAT RED SPOT

**Categories:** Planets and Planetology; The Jovian System

*High-resolution data have been obtained concerning the nature of Jupiter's Red Spot, a weather system with horizontal dimensions comparable to the diameter of Earth and monitored behavior spanning centuries.*

## Overview

Excluding Jupiter's general east-west belt and zone pattern, the Great Red Spot is the most obvious, persistent, and continuously observed feature of Jupiter's visible cloud deck. Centered at about 20° south latitude, it spans about 14,000 kilometers in the north-south direction and about 26,000 kilometers in the east-west direction. Compared with the diameter of Earth, the Red Spot is a huge cloud structure large enough to span two Earths.

This well-defined oval feature has raised considerable curiosity ever since its discovery, which many credit to original observations made by Giovanni Cassini or Robert Hooke in the seventeenth century. Coordinated reports by the British Astronomical Society and original drawings maintained in the Royal Astronomical Society library collections in London definitively established that the Red Spot has been present since at least 1830. Earlier scattered reports of pink spots in the atmosphere of Jupiter extend back several centuries to the era of earliest efforts to improve the resolution of simple telescopes.

Realizing that they were seeing an opaque cloud deck and that the surface of Jupiter was never visible, some observers suggested that the Red Spot was caused by interference between an elevated surface feature and the prevailing winds. In the early 1960's Raymond Hide proposed a model for the driving mechanism. If this model had withstood scrutiny, it would have permitted scientists to calculate the planet's surface rotation rate and to interpret other cloud motions by extrapolating from this rate. Careful examination of measured periods of rotation indicated, however, that Hide's model was inconsistent with available data. The Red Spot circles Jupiter at an almost unvarying speed for a period of twenty to fifty years. At the end of each period, it suffers an acceleration or deceleration which occurs over a period of weeks. After this adjustment period, the Red Spot continues to circle at its new speed. This behavior indicates that it cannot be the result of an upwardly propagating disturbance above an elevated region on a hidden surface. Measured positions indicate that there is no rate at which the interior of the planet could rotate that would not force the Red Spot to drift freely either east or west within the atmosphere, yet historical observations indicate that the spot is trapped in the prevailing east-west wind pattern and is not free to move north or south like weather systems in Earth's mid-latitudinal regions.

Along the southern edge of Jupiter's equatorial zone, winds blow eastward at 150 meters per second. West-to-east zonal winds decrease poleward, until at 17.5° south latitude they are moving westward at 70 meters per second. From 17.5° to 24.5° latitude, winds increase eastward to a maximum of nearly 60 meters per second. From 24.5° to 50°, winds alternate eastward and westward. This alternating east-west wind pattern, with four cycles between the equator and 50° south latitude, generates significant latitudinal wind shear. If local heating occurs below the cloud deck, causing the atmosphere to rise and clouds to form, maintaining a long-lived cloud system in the presence of strong horizontal shear would require the cloud to rotate about its center. If the cloud rotates in the same sense as the local horizontal shear, it can deflect the prevailing winds about its perimeter. The Red Spot displays this behavior. Not only is it trapped between westward winds at 17.5° and eastward winds at 24.5°, it also deflects westward wind flow around its equatorward perimeter, creating a large indentation, or hollow, in the poleward side of the dark adjacent belt. Other, smaller oval cloud systems are associated with the more poleward wind-shear regions. Three white oval cloud systems, noted in 1938, are located near 29° south latitude. The east-west dimension of each of these systems is about 12,000 kilometers. A series of smaller ovals circle the planet near 37° south latitude. Morphologically, the Red Spot is not unique. However, it is the largest example of a type of cloud system common to the southern hemisphere of the planet.

The Red Spot is notable not only in size but also in coloration. Jupiter's other oval clouds are white, indicating that their cloud decks are composed of highly reflective ammonia ices. When visible red and infrared reflection from the Red Spot is analyzed, data indicate ammonia ice is present there as well. The Red Spot has additional trace constituents in its cloud deck that are strong absorbers of ultraviolet, violet, and blue wavelengths. This gives the Red Spot its unique color. Small, short-lived ovals that form at similar latitudes in the northern hemisphere also absorb ultraviolet and blue light. This suggests that these absorbers are carried upward from below. Also the rate of vertical motion or the depth to which the convective

*The Great Red Spot is seen in this Voyager 2 image, appearing as the large, eye-shaped oval to the right of center.* (NASA/JPL)

until 1938. Then the belt just south of the South Tropical Zone underwent a major disruption resulting in greatly increased reflectivity of the belt and the formation of three white ovals. In the early 1930's, the Red Spot drifted at a rate similar to that seen prior to 1878. After formation of these ovals, the Red Spot decelerated to its slowest drift rate ever observed. Since 1962, the Red Spot has been drifting at a rate similar to that of the 1878-1901 period.

This constitutes evidence that the Red Spot interacts with its surroundings and that variations in local temperature, pressure, and wind patterns occur. Even so, the entire range of variation in average Red Spot motion, with the average velocity derived from the annual longitudinal displacement of the Spot relative to the rotation rate of radio noise, lies between −4.4 and −0.6 meters per second. Although this variation is small when compared to daily wind speeds at midlatitudes on Earth, an annual increase of 2 meters per second in wind speed results in an eastward displacement of about 63,000 kilometers, or about two and a half times the Spot's length.

That the Spot's recovery from a given acceleration or deceleration takes years is expected. A body's heat loss rate depends heavily on the relative temperatures of the body and its surroundings. The Jovian cloud deck temperature is approximately 153 kelvins. Thus, the rate of heat loss to black sky is relatively slow. It is logical that once an excess amount of heat has been inserted into the atmosphere, it will be several years before the atmosphere returns to its previous state. One basic question that atmospheric scientists wanted to answer concerned the nature of acceleration mechanisms. Ground-based observations indicated that these events occurred over short time intervals.

During the 1960's and early 1970's, Elmer Reese made many detailed measurements using photographs of Jupiter. One result of this work was a measurement of Red Spot rotation. The Red Spot rotates counterclockwise, completing one rotation every twelve days. A feature in Earth's atmosphere with this behavior would have air rising in the center, flowing outward at the cloud top,

motion reaches permits transport not present at the top of the cloud deck in storms located at more poleward latitudes.

A trip to a mountaintop on Earth's surface makes it clear that lower elevations of Earth's atmosphere are compressed. Jupiter's atmosphere must behave similarly. In order for the Red Spot to behave as an isolated system, its vertical dimension must be small in relation to its horizontal extent. Comparisons of the Red Spot with a hurricane are inappropriate. The Red Spot is a giant rotating cloud system, trapped in the prevailing winds. Reflectivity and degree of redness vary with time; still, deflection of the westward jet around the equatorward side of the Red Spot is always visible.

In 1878 the Red Spot suffered a deceleration. The surrounding cloud deck became highly reflective and white; however, the Red Spot remained dark and red. This sharp contrast made many casual observers aware of the phenomenon. In 1901, a disturbance occurred in the South Tropical Zone, the white band south of the Red Spot. This event appeared to be a major weather disturbance that moved eastward and caught up with and then passed the Red Spot, thereby accelerating the Spot. This continued and descending around the perimeter. Measurement of divergent flow was one goal of the two Voyager spacecraft. Superimposed on the drift is an oscillatory motion of the whole feature, speeding up and slowing down so that its velocity oscillates every ninety days, causing the Spot to shift back and forth about 900 kilometers relative to its average path. This behavior apparently results from some natural period of response of the system to its surroundings.

Because the Red Spot had already been subjected to much scrutiny, planetary scientists eagerly looked forward in turn to obtaining high-resolution data from Pioneer and Voyager flybys in the 1970's and subsequently from the Galileo orbiter between December, 1995, and September, 2003. The Galileo spacecraft's science mission centered on investigations of the planet's particles and fields, various satellites, and general atmospheric dynamics; the latter included an atmospheric probe, but that heavily instrumented payload was not flown into the Red Spot. Additional high-resolution images of the Red Spot were obtained from the Earth-orbiting Hubble Space Telescope and during flybys of probes passing through the Jovian system to gain a gravity asset from Jupiter. Those flybys provided opportunities for controllers to test scientific packages on those spacecraft. The Cassini orbiter in December, 2000, on its way to the Saturn system, obtained images of the Red Spot superior to those from the Voyager spacecraft, even though its closest approach to Jupiter was considerably farther out. New Horizons in February, 2007, on its way to the Pluto-Charon system, obtained high-resolution images of Jupiter. It also verified that observations from ground-based telescopes and the Hubble Space Telescope had indeed recorded a lightening phase of the Red Spot beginning in 2006. In addition, New Horizons studied a relatively new storm farther south of the Great Red Spot, one referred to as the Little Red Spot, a much smaller version of its long-lived sibling.

The Little Red Spot started as one of three white storms that formed in the 1940's. Two of those merged in 1998, and the resulting white spot merged with the third in 2000. This feature then demonstrated a trend toward reddening in late 2005, and after 2006 it was referred to as the Little Red Spot. This storm continued to grow in wind speed and in reddish hue, providing a marvelous opportunity for in-depth study by the contemporary technology of the passing New Horizons spacecraft. Comparisons of the Little and Great red spots were expected to shed light on the complex atmospheric physics raging within these large-scale and unique storms. That research continues.

## Knowledge Gained

Both Pioneer spacecraft were spin-stabilized, so that the planet swept past their instruments' fields of view. This design feature placed strong constraints on the type of instrumentation that could be implemented. The imaging experiment utilized a scanning camera that sampled one point on the planet at a time. Images were constructed by scanning the planet row by row as it passed the field of view. This method limited the ultimate resolution of the data and severely curtailed the number of images that could be recorded. Nevertheless, Pioneer data were highly useful to astronomers. Tom Gehrels and his team observed Jupiter at a time when the belt adjacent to the Red Spot was highly reflective and white. The Red Spot was quite dark. Comparison with descriptions of the Red Spot in 1878 indicated that reflectivity of the Spot and its surroundings at the time of both Pioneer encounters was highly similar to its condition approximately a century earlier.

That Red Spot behavior continued until July, 1975, when a bright white cloud appeared west of the Spot that expanded rapidly and sheared out in the zonal winds. Considerable turbulence accompanied this event, and within a few weeks the Red Spot and belt had changed significantly. Material from the disturbance encountered the Red Spot from the west and formed a large white mass of clouds to the west of the Spot. Turbulent cloud masses also spilled into the westward wind jet along the south side of the belt. This material was carried around the planet and approached the Red Spot from the east side. The contrast of the Red Spot decreased as it became whiter. Historically, Red Spot lightening has been fairly common. It appears that increased turbulence and vertical mixing in the belt that lies to the equatorward side of the Red Spot carry ammonia ices into the Spot. The Spot retained this appearance from 1975 to 1987 and was observed at high resolution in this condition by the Voyager spacecraft.

Voyagers 1 and 2 arrived at Jupiter in March and July of 1979. The two spacecraft were equipped with two television cameras each; one had a wide field of view and the other focused on a smaller field with higher resolution. The two were boresighted; thus, simultaneous views allowed detailed sampling, while defining the direction that the cameras were pointing. Ultraviolet and infrared spectrometers and a photopolarimeter allowed observations as a function of wavelength. Red Spot images with higher spatial resolution than could be obtained from Earth were taken over a period of three months with each spacecraft.

The Galileo probe began collecting visible light and near-infrared images in 1996. The Hubble Space Telescope, Cassini orbiter, and New Horizons spacecraft also supplied data.

Ultimately, high-resolution sequences requiring as many as twenty-seven narrow-angle camera frames to map the Red Spot were executed. Resulting data revealed details of the flow pattern around the Red Spot. Winds were deflected around the Spot; small ammonia ice clouds, however, were observed to pass around the equatorward edge and to continue around the western cusp and along the feature's southern edge. When these clouds reached the Spot's southeast corner, they moved into the Red Spot and were sheared apart to form a high-velocity collar inside the Red Spot. Jim Mitchell, Reta Beebe, Andrew Ingersoll, and others analyzed the flow within the Red Spot and the white ovals. Velocities of rotation about the Spot's center as high as 150 meters per second were measured in the outer third of the feature. In the inner half of the Spot, reflectivity was lower, and motion of the cloud deck was small and random. No outward flow from the center toward the perimeter of the Spot was detected. When infrared data from the Red Spot and one of the white ovals were compared, no difference in absorption as a function of color could be detected; thus, the infrared data offered no clues to the identity of the ultraviolet absorber. This finding was not unexpected, because it was known that the ammonia ice would tend to dominate in the infrared.

## Context

High-resolution spacecraft imagery has been combined with long-term, lower-resolution ground-based photography in an effort to understand the Red Spot's nature. Apparent motion of planetary atmospheric features can be attributed to mass motion when material is physically translated in the zonal wind or to wave motion. In the case of wave motion, variations in local pressure and temperature introduced by the wave cause local condensation or evaporation. Many of the small-scale patterns that add beauty to Earth's water clouds are of the second type. Thus, in an effort to elucidate the Red Spot's nature, models that consider different types of wave structures have been constructed. Not all wave structures are a series of oscillations with equal amplitudes traveling through space. By varying modeled environmental conditions

*A near-infrared mosaic of Jupiter's Great Red Spot from images taken by the Galileo spacecraft.* (NASA/JPL)

within which the wave is formed, various researchers, including Tony Maxworthy, Andrew Ingersoll, and Gareth Williams, have investigated the characteristics of waves and related them to the morphology of the Red Spot.

In order to construct a realistic model of the Red Spot, information concerning the manner with which the zonal winds change with depth as well as with latitude is necessary. Because all required parameters are not available, a series of models must be constructed. Peter Read and others have attempted to shed light on atmospheric flow around the Red Spot by constructing cylindrical tanks of rotating fluids, within which they generate closed eddies that have characteristics similar to those of the Red Spot. High-resolution spacecraft data have provided astronomers with a wealth of information. Data have stimulated computer analysis and the gathering of additional understanding through a great deal of observation and experimentation.

It is not clear why the well-formed oval clouds in Jupiter's atmosphere preferentially form in the southern hemisphere. The fact that they are very long-lived is, however, consistent with their being large, closed eddies rotating in the local wind shear. Little is known concerning the rate of vertical motion associated with these features or the depth to which they extend below the cloud deck. Interplanetary spacecraft will continue to provide high-resolution data for researchers struggling to define the nature of Jupiter's Great Red Spot. In the planning phase for perhaps the second decade of the twenty-first century, a proposed Jupiter Icy Moons Orbiter would also provide prolonged observation of Jupiter's atmosphere, including the dynamics and evolution of the Great Red Spot.

The variability of Jupiter's atmosphere, as well as its ability to sustain prolonged features, was illustrated when a number of smaller red spots broke out beginning in early 2006. The first small spot continued into 2008, when in May a third spot appeared, this one located in the southern hemisphere farther south in longitude than the Great Red Spot. Both little red spots moved in a way that led scientists to expect them to merge. However, the influence of the Great Red Spot held the potential to push both little red spots to the side. Coordinated observations of these small storms from the Hubble Space Telescope and ground-based telescopes in visible and near-infrared light suggested that these little red spots started as white storms and then assumed a reddish color. Jupiter appeared to be undergoing a global climate change in which the equator was warming and the south pole was cooling. As a result,

jet streams in the southern hemisphere were destabilized such that new storms could be generated.

Near the end of May, 2008, the first of two little Red Spots in the planet's southern hemisphere began to grow both in size and in wind speed. This new rival to the Great Red Spot developed winds of 172 meters per second, very nearly the same speed as winds in the Great Red Spot. It remained to be seen if this Little Red Spot would continue to grow independently or be swallowed up by the extremely long-lived Great Red Spot. Its general motion after its increase in size and wind speed was toward the Great Red Spot.

*Reta Beebe and David G. Fisher*

## Further Reading

Beatty, J. Kelly, Carolyn Collins Petersen, and Andrew Chaikin, eds. *The New Solar System.* 4th ed. Cambridge, Mass.: Sky, 1999. The editors have assembled a collection of articles written by top researchers in the field as a survey of then-current knowledge of the solar system bodies. One chapter places the Red Spot in its atmospheric context. Includes many illustrations.

Consolmagno, Guy. *Worlds Apart: A Textbook in Planetary Sciences.* Englewood Cliffs, N.J.: Prentice Hall, 1994. A text accessible to college-level science students, using low-level mathematics as well as integral calculus where required. Demonstrates how the area of planetary science progresses by questioning previous understandings in the light of new observations.

Fimmel, Richard O., James Van Allen, and Eric Burgess. *Pioneer: First to Jupiter, Saturn, and Beyond.* NASA SP-446. Washington, D.C.: Government Printing Office, 1980. This detailed overview of the Pioneer mission includes reproductions of the Red Spot images obtained by the spin-scan camera.

Fischer, Daniel. *Mission Jupiter: The Spectacular Journey of the Galileo Spacecraft.* New York: Copernicus Books, 2001. A detailed overview of Galileo's investigations of the Jupiter system. Provides numerous full-color images taken from Galileo and other spacecraft.

Gehrels, Tom, ed. *Jupiter.* Tucson: University of Arizona Press, 1976. This collection of scientific essays on all aspects of Jupiter reflects the pre-Voyager state of knowledge of the planet. Still, its copious documentation will point the interested reader to the journals and other sources that present later findings. A challenging text; requires some background in astronomy.

Harland, David M. *Jupiter Odyssey: The Story of NASA's Galileo Mission.* New York: Springer, 2000. Provides a detailed account of all major Galileo spacecraft operations and the mission's scientific investigations of the Jupiter system. Filled with technical diagrams and images from Galileo and other spacecraft.

Hartmann, William K. *Moons and Planets*. 5th ed. Belmont, Calif.: Thomson Brooks/Cole, 2005. An excellent text for a course on planetary science, accessible to advanced high school students and undergraduates alike. Covers the entire solar system and includes much about the Red Spot. Takes a comparative planetology approach rather than including separate chapters on individual planets in the solar system.

Hunt, Garry E., and Patrick Moore. *Jupiter*. New York: Rand McNally, 1981. Reviews the Voyager mission and describes the Jovian system. Features numerous color photographs and illustrations of the Red Spot. Accessible to the general reader.

Irwin, Patrick G. J. *Giant Planets of Our Solar System: An Introduction*. 2d ed. New York: Springer, 2006. Suitable as a textbook for upper-level college courses in planetary science. Focuses on Jupiter, Saturn, Uranus, Neptune, and their satellites, rings, and magnetic fields. Filled with figures and photographs.

Morrison, David, and Jane Samz. *Voyager to Jupiter*. NASA SP-439. Washington, D.C.: Government Printing Office, 1980. A description of the events leading up to the Voyager missions. Communicates the excitement experienced by scientists as they received close-up views of the Red Spot. Contains many color reproductions.

Peek, Bertrand M. *The Planet Jupiter*. London: Macmillan, 1958. A classic detailed summary of ground-based observations recorded by members of the British Astronomical Association. This book provides an overview of time-dependent aspects of the behavior of the Red Spot.

# JUPITER'S INTERIOR

**Categories:** Planets and Planetology; The Jovian System

*Jupiter, the largest planet in the solar system, is often described as a "gas giant" planet. However, the interior structure of Jupiter is far more complex than a big ball of gas. These interior characteristics are also responsible for many of the planet's observed properties.*

## Overview

Jupiter is the largest planet in the solar system, with a mean diameter of about 138,000 kilometers. Jupiter's mass, about 320 times more than Earth's mass. This is greater than that of the rest of the planets combined. However, most of Jupiter's mass is made up of the two lightest elements in the universe: hydrogen and helium. Jupiter is believed to be formed directly from the disk of material swirling together to make the Sun. Thus it is not a surprise to find that Jupiter's composition is very similar to that of the Sun (a star), which is also mostly hydrogen and helium. However, there are significant differences between the structure of Jupiter and that of a star, or even a failed star such as a brown dwarf. The interior structure of Jupiter has not been directly observed. It can only be inferred through mathematical modeling based on observations of the planet. Jupiter does not have a solid surface on which a spacecraft can land, and cloud layers in Jupiter's upper atmosphere shield the interior of the planet from view.

Jupiter is observed to have an oblateness of 0.065 (its equatorial diameter is 6.5 percent greater than the pole-to-pole diameter). This observation constrains the size of any solid core that the planet may have. Mathematical models suggest that Jupiter has a rocky core, with perhaps a nickel-iron inner core, of mass somewhere between 8 and 15 times the mass of the Earth. Due to Jupiter's large mass, though, that is only between 2.5 to 4.7 percent of Jupiter's mass. Despite the core being several times the mass of Earth, the extreme pressure inside Jupiter, nearly 70 million atmospheres, compresses the core to a size on the order of that of Earth. The temperature at the core of Jupiter is believed to be perhaps 22,000 kelvins. It is not known if the core is solid or liquid.

Because Jupiter is in a region of the solar system in which a large number of icy bodies exist, a great number of these bodies must have impacted Jupiter over its history, starting at its formation. These ices, being heavier than the hydrogen and helium that make up the bulk of Jupiter, would have settled toward the deep interior of the planet. Planetary scientists believe that there may exist a layer of this material perhaps 3,000 kilometers thick on top of the core. The ices include frozen ammonia and methane rather than just water ice. The temperature and pressure deep inside Jupiter would ensure that this material is in a liquid state, though, rather than frozen. Therefore, the term "liquid ices" is often used to describe this layer. Great pressure at this depth would make this material behave in ways quite different from the way the same material would behave on Earth.

The bulk of Jupiter is composed of hydrogen and helium. The outermost parts of Jupiter are gaseous, with clouds in the upper 100 kilometers. At great depths inside the planet, the pressure becomes great enough to compress these gases into a liquid state. That pressure is reached at a depth of about 1,000 kilometers below the cloud tops. However, there is no vast ocean of liquid hydrogen under Jovian skies the way that Earth's water collects in oceans. On Jupiter, in fact, there is no clear boundary between the gaseous atmosphere and the liquid interior, because the temperature and pressure inside Jupiter are well in excess of hydrogen's critical point. Beyond the critical point of a substance, there ceases to be a definite phase transition between liquid and gas. Rather, the material takes on a state known as a supercritical fluid. At greater altitudes, the hydrogen in Jupiter is clearly gaseous. At much lower levels, it definitely has more liquid properties, but there is no obvious depth at which the hydrogen becomes liquid. Instead, with increasing depth, the hydrogen becomes more and more like liquid. Though Jupiter is called a "gas giant" planet, the majority of the planet's composition is actually liquid.

At sufficient pressure and temperature, hydrogen takes on metallic properties. That means that it conducts heat and electricity like any other element on the left-hand column of the periodic table of elements. These conditions are met in Jupiter below a depth of about 7,000 kilometers below the planet's cloud tops. Liquid metallic hydrogen exists from that depth all the way down to the liquid ices at the core. That means that the bulk of Jupiter's mass is in a mantle composed of helium and liquid metallic hydrogen, possibly comprising about two-thirds of the planet.

Jupiter has the strongest magnetic field of any planet in the solar system. At its equator, Jupiter's magnetic field is nearly fourteen times stronger than Earth's magnetic field. Planetary magnetic fields are believed to be created by magnetohydrodynamics in a planet's interior. A dynamo model of planetary magnetic fields shows that a suitable conductor moving in a magnetic field can regenerate that magnetic field, producing a long-lived magnetic field. However, this dynamo effect requires a highly conducting fluid in order to operate. It may be possible for part of Jupiter's core to have a liquid iron region, but that would be not be large enough to account for Jupiter's magnetic field. Rather, Jupiter's magnetic field originates primarily in its liquid metallic hydrogen mantle.

Unlike Earth, Jupiter radiates nearly twice as much energy as it gets from the Sun. This surplus energy is produced by kelvin-Helmholtz contraction. When planets form, a large amount of gravitation energy is released as the materials that form the planet come together. For fluid bodies such as Jupiter, as they radiate thermal energy into space, they contract somewhat. This contraction then compresses the material making up the planet, heating it further. Other gas giant planets besides Jupiter probably also had kelvin-Helmholtz contraction after they formed, but they have long since stabilized at a point where such contraction no longer is a major source of thermal energy. Jupiter is at nearly the perfect mass to extend kelvin-Helmholtz contraction to the longest time possible. If Jupiter had more mass, then it would have compressed faster, until it reached a point of maximum compression. If Jupiter had

## Jupiter Compared with Earth

|  | Jupiter | Earth |
|---|---|---|
| Mass ($10^{24}$ kg) | 1898.6 | 5.9742 |
| Volume ($10^{10}$ km³) | 143,128 | 108.321 |
| Equatorial radius (km) | 71,492 | 6378.1 |
| Ellipticity (oblateness) | 0.06487 | 0.00335 |
| Mean density (kg/m³) | 1,326 | 5,515 |
| Surface gravity (m/s²) | 23.12 | 9.78 |
| Surface temperature (Celsius) | −140 | −88 to +48 |
| Satellites | 63 | 1 |
| Mean distance from Sun millions of km (mi) | 779 (483) | 150 (93) |
| Rotational period (hrs) | 9.9250 | 23.93 |
| Orbital period | 11.86 yrs | 365.25 days |

*Source:* Data are from the National Aeronautics and Space Administration/Goddard Space Flight Center, National Space Science Data Center.

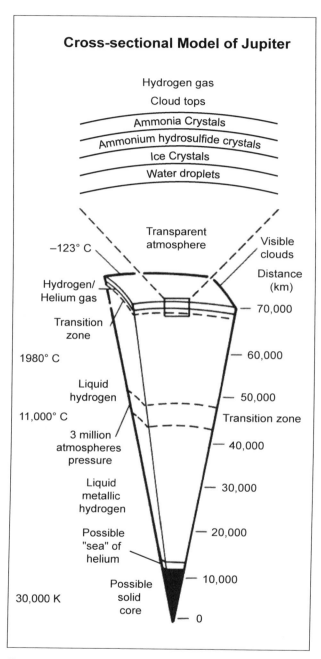

### Cross-sectional Model of Jupiter

Hydrogen gas
Cloud tops
Ammonia Crystals
Ammonium hydrosulfide crystals
Ice Crystals
Water droplets

Transparent atmosphere
−123° C
Visible clouds
Distance (km)
Hydrogen/ Helium gas
70,000
Transition zone
60,000
1980° C
Liquid hydrogen
50,000
11,000° C
Transition zone
3 million atmospheres pressure
40,000
Liquid metallic hydrogen
30,000
Possible "sea" of helium
20,000
10,000
30,000 K
Possible solid core
0

*Source:* David Morrison and Jane Samz. Voyage to Jupiter. NASA SP-439. Washington, D.C.: National Aeronautics and Space Administration, 1980, p. 4.

less mass, like Saturn, then it would have not had sufficient gravity to keep contracting for as long as it has.

## Knowledge Gained

Much has been learned about Jupiter through observations from Earth. The interior of the planet, of course, cannot be directly measured. Understanding the nature of matter has allowed astrophysicists to make theoretical models of Jupiter's interior; however, it took measurements by spacecraft sent to Jupiter to actually begin to learn more about that planet's interior structure.

Since Jupiter's magnetic field is produced in the planet's mantle, it rotates with the planet's interior. Studies of the magnetic field show that Jupiter's interior rotates once every 9 hours, 55 minutes, and 30 seconds, somewhat more slowly than the rate of rotation of the cloud tops near the planet's equator. Until spacecraft were able to approach Jupiter to measure its magnetic field, astronomers could only guess at its interior rotational period. Jupiter's huge liquid metallic hydrogen layer produces a magnetic field that is so powerful that Jupiter has a magnetic field stronger than any other planet in the solar system, and Jupiter's magnetosphere is the largest of any of the planets. The existence of Jupiter's powerful magnetic field provided evidence of the metallic nature of hydrogen well before it was produced in the laboratory.

To date, most of the extrasolar planets discovered have been gas giant planets. As the largest gas giant planet in the solar system, studies of Jupiter help to reveal the nature of these planets. Gas giant planets can have masses greater or less than that of Jupiter. However, in size, Jupiter is about as large as a gas giant planet can be. If it had much more mass, gravity would compress it to a smaller volume. Less mass would, of course, make a smaller planet, but its lower gravity would compress the planet less. For example, Saturn has almost 30 percent of Jupiter's mass but more than 80 percent of Jupiter's diameter and more than 55 percent of Jupiter's volume.

Studies of Jupiter's composition have led astronomers to believe that Jupiter may have formed somewhat farther from the Sun than the distance at which it currently orbits. Interactions with other planets, notably Saturn, could have caused Jupiter to migrate inward. This planetary migration also explains observations of extrasolar gas giant planets that appear much closer to their stars than can be explained through current understanding of planetary formation.

In late 2008 a research team reported the results of computer simulations based on the properties of hydrogen-helium mixtures exposed to the extreme conditions of temperature and pressure deep inside Jupiter. Their computer simulation also incorporated a core accretion model. In a paper published in the November 20, 2008, issue of *Astrophysical Journal Letters*, these researchers presented an argument that Jupiter's core could be twice as big as previously thought. Their simulation predicted a rocky core perhaps amounting to 5 percent of Jupiter's total mass, making the rocky core equivalent to 14 to 18 Earth masses. That core would have layers of metals, rocky material, methane ice, water ice, and ammonia ice. Like Earth's core, the very center of Jupiter's core would be composed of iron and nickel. This computer simulation could be applied to attempts to understand the cores of the other gas giant planets as well. However, being a computer model, additional observational data and further analysis would be needed before this intriguing claim could achieve complete acceptance from the planetary science community.

## Context

Jupiter and Saturn, the two largest planets in the solar system, probably share a common origin and similar structure. Both formed largely from material that was coming together to form the Sun. Thus, studies of these two worlds allow astronomers to learn more about conditions in the early solar system. Understanding these planets helps astronomers to understand how other planets, including the "rocky" planets such as Earth, form. Jupiter also seems to be similar to exoplanets that have formed around other stars and thus a sort of laboratory for understanding those extrasolar planets.

However, Jupiter is still nearly 600 million kilometers away from Earth, even at its closest approach. Thus, detailed studies have required visits by spacecraft. In total, seven spacecraft have studied Jupiter. Launched in the early 1970's, Pioneer 10 and Pioneer 11, followed in the late 1970's by Voyager 1 and Voyager 2, eventually flew past Jupiter on their way to the outer solar system. Early in 2007, the New Horizons spacecraft flew past Jupiter on its way to Pluto and the Kuiper Belt. The Cassini spacecraft passed by Jupiter December 30, 2000, on its way to Saturn. The Ulysses spacecraft flew by Jupiter in February, 1992, using that planet's gravity to send it into an orbit that permitted it to study the Sun's polar regions. All of these spacecraft studied Jupiter as they went past. The Galileo

spacecraft, however, was sent specifically to study Jupiter, orbiting that planet from 1995 to 2003. Galileo also sent an atmospheric probe into Jupiter, the only probe to enter the atmosphere of any of the gas giant planets.

Jupiter is the best studied of the gas giant planets, therefore, but it still holds many mysteries. Its interior must still be investigated through inferences from observations of Jupiter's exterior and of the planet's magnetic field. Debate continues over the exact nature of the planet's interior structure as astronomers pursue additional studies to develop a detailed understanding of this world.

*Raymond D. Benge, Jr.*

## Further Reading

Bagenal, Fran, Timothy E. Dowling, and William B. McKinnon, eds. *Jupiter: The Planet, Satellites, and Magnetosphere.* New York: Cambridge, 2004. A collection of papers on the current scientific understanding of the planet Jupiter, including a chapter on Jupiter's interior.

Corfield, Richard. *Lives of the Planets: A Natural History of the Solar System.* New York: Basic Books, 2007. This book focuses less on the planets themselves than on the history of planetary exploration and the spacecraft that have studied the planets.

Fischer, Daniel. *Mission Jupiter: The Spectacular Journey of the Galileo Spacecraft.* New York: Copernicus Books, 2001. A good overview of the Galileo mission to Jupiter. Some of the initial findings are given, along with many color photographs of the planet and its moons.

Freedman, Roger A., and William J. Kaufmann III. *Universe.* 8th ed. New York: W. H. Freeman, 2008. An excellent college-level introductory astronomy textbook. An entire chapter is devoted to Jupiter and Saturn.

Irwin, Patrick. *Giant Planets of the Solar System: An Introduction.* New York: Springer, 2006. An overview of all four gas giants in the solar system, this text is written at the level of advanced students. It covers all aspects of the planets, including theories of formation, and has a very good bibliography.

McAnally, John W. *Jupiter and How to Observe It.* New York: Springer, 2008. Intended for amateur astronomers, this book focuses on observations of Jupiter, but it also includes information on the planet itself.

# JUPITER'S MAGNETIC FIELD AND RADIATION BELTS

**Categories:** Planets and Planetology; The Jovian System

*An understanding of Jupiter's magnetic field has proved vital to furthering comprehension of this enormous planet's singular structure, and such knowledge also enriches Earth, planetary, and solar-system science.*

## Overview

Since 1610, when Galileo focused his telescope on Jupiter and four of its satellites, this immense planet, orbiting 628 million kilometers from Earth at its closest approach, has received much attention from astronomers. Jupiter's mean distance from the Sun is 5.2 times the mean distance between the Earth and the Sun, the latter being known as an astronomical unit (AU). After the Sun, Jupiter is the largest object in the solar system, possessing a mass 318 times greater than Earth's and a diameter 11 times longer. Its volume is thirteen hundred times that of Earth. Because of its prominence, many of Jupiter's basic characteristics were ascertained centuries ago. Sir Isaac Newton accurately calculated its mass and density, for example. Its radius, diameter, rate of rotation, chemical composition, and singular surface features have similarly been under study for centuries. However, this body, which is neither a star like the Sun nor a terrestrial planet like Earth, has retained certain mysteries, many of which have to do with the composition of Jupiter's interior and the origins and behavior of its magnetic field and radiation environment.

During the 1950's and 1960's, prior to investigations by uncrewed spacecraft, radio astronomers gathered approximate data on Jupiter's magnetosphere, that zone of powerful magnetic influence that surrounds the planet. In 1955, Bernard Burke and Kenneth Franklin, both radio astronomers, found evidence that Jupiter's magnetosphere was a source of nonthermal radio activity (in contrast to the thermal radiation emitted by all objects with temperatures above absolute zero) at a frequency of 22.2 megahertz (MHz). Other astronomers later noted that this radio activity occurred, if not continuously, at least in a patterned way, at the same point in the planet's rotation. Such emissions distinguished Jupiter from the other planets and raised provocative questions about the validity of previous theories of radiation.

Several years later, additional radio bursts were picked up by Earth-based radio telescopes in a different portion of the radio spectrum (300 to 3,000 MHz). Unlike the emissions detected earlier, which originated at the planet's surface, these decametric radiations emanated from within Jupiter's toroidal region—a region encircling the planet, tilted about 10° from its equatorial belt and extending about 286,400 kilometers from it into space. Within this region and on both sides of the planet at about 140,000 kilometers from its surface there are two "hot spots," areas of intense radiation activity, which evoked considerable scientific curiosity.

In 1959, two astronomers, Frank Drake and S. Hvatum, identified the source of this toroidal radio activity as synchrotron radiation. Atomic particles within the region were being accelerated to very high speeds by a powerful magnetic field and by changes in the frequency of the electric field. There has been a growing consensus among astronomers that these high-energy particles, electrons that came from the Sun, have been trapped by Jupiter's external magnetic field, which is 19,000 times stronger than Earth's magnetic field. Forming radiation belts around the planet, these high-energy particles, moving at high velocities, may produce radio emissions when they strike the top of Jupiter's atmosphere.

Growing scientific curiosity about these prodigious emissions led to Jupiter's magnetosphere becoming a subject for investigation during the flybys of Pioneers 10 and 11. Pioneer 10, launched in March, 1972, flew to within 130,000 kilometers of Jupiter on December 3, 1973, securing remarkable photographs of Jovian cloud tops and measuring Jupiter's radiation belts before sailing farther into space. Pioneer 11, also referred to as Pioneer-Saturn, was launched in April, 1973. On December 2, 1974, it flew to within 42,000 kilometers of Jupiter, photographing the vast but previously unstudied Jovian polar regions. Among other measurements, Pioneer 11 obtained fresh data on Jupiter's complex magnetic field. The massive flow of data from the Pioneer flights greatly contributed to more precise assessments of the configurations of Jovian radiation belts and the magnetic field, as well as of the extent of the magnetosphere and the distribution of energetic electrons and protons within the planet's interior. This information, in turn, encouraged fresh ideas about Jupiter's structure and rotation. In the mid-1970's, for example, John D. Anderson of the Jet Propulsion Laboratory in Pasadena, California, and William B. Hubbard of the University of Arizona made use of Pioneer data to devise a new model of Jupiter's internal structure that was consistent with knowledge about both its gravitational field and its magnetic field.

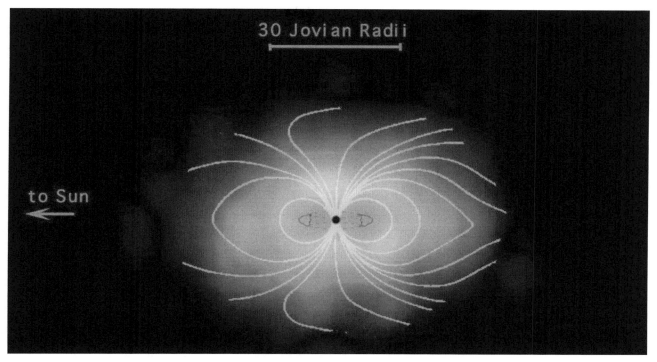

*Jupiter's huge magnetosphere, which covers more space than the Sun, as imaged by instruments aboard the Cassini spacecraft on December 30, 2000.* (NASA/JPL/Johns Hopkins University Applied Physics Laboratory)

Anderson and Hubbard proposed that beneath the dense and apparently chaotic Jovian atmosphere lies a thick layer of liquid molecular hydrogen, beneath which an even thicker layer of liquid metallic hydrogen exists. The heart of the planet, they proposed, consists of a small, rocky core of iron and silicates heated to temperatures of nearly 30,000 kelvins. Although the presence of such a core could never be proved by gravitational studies, its existence was plausibly deduced from the assumption that if its composition were similar to the Sun's, it should contain some measure of the same elements. The liquid metallic hydrogen layer presumably extends 46,000 kilometers out from the core, is heated to 11,000 kelvins, and is compressed under a pressure of 3 million Earth atmospheres. This layer cannot as yet be experimentally modeled in bulk in a laboratory, yet the construct is plausible: Metallic hydrogen has been created in small amounts in the laboratory. The first to do it were researchers at Lawrence Livermore Laboratory in 1996. In a liquid metallic state, hydrogen molecules break down into atoms and become electrical conductors.

Information from the Pioneer flybys of Jupiter also led to revisions of the hypothesis that the planet's excessive radiation of heat results either from radioactivity or from heat generated by gravitational contraction of the largely gaseous mass of which Jupiter is composed. Since it now appeared that Jupiter was a liquid body (and liquids are all but incompressible), it seemed likely that its excess radiated heat was merely a residue of the heat generated when the planet coalesced from the solar nebula. The implication would be that the planet's original thermal energy is continuously finding its way to the surface and, in the process, creating convection currents, the rising of hot gases or liquids and the downward movement of cold liquids or gases in the planet's interior. Such grand-scale convection currents, as described by John H. Wolfe, who served as chief scientist for the Pioneer missions to Jupiter, could constitute a mechanism for generating the Jovian magnetic field. As primordial heat rose through Jupiter's liquid metallic hydrogen, stirring it, the Coriolis force affected the resulting convection currents. The Coriolis force arises from planetary rotation and deflects other forces in motion—depending on whether they are north or south of the body's equator—either to the right or to the left, much as a person walking across a moving carousel appears thrown off course as seen in an inertial reference frame. Deflected convection currents in such circumstances would set up loops of electric current, which could and may create a magnetic field.

The magnetosphere of Jupiter was determined to expand and contract under pressure from the solar wind. Where an equilibrium existed between magnetic forces and the solar wind's incident stream of charged particles, a planetary magnetopause developed. Pioneer data confirmed the magnetopause to be as far out from the Jovian atmosphere as 7,135,000 kilometers and as close to its atmospheric layer as 3,565,000 kilometers. Just as there is a bow shock wave, there is also a magnetotail, which, much like a ship's wake, extends several hundred kilometers behind Jupiter in a direction away from the Sun.

Like Earth's magnetic field, the Jovian field is dipolar, but its magnetic axis is tilted between 9.5° and 10.8° from the planet's rotational axis, a displacement of about 7,000 kilometers from Jupiter's center. The strength of the magnetic field, as measured at Jupiter's cloud tops, varies from 3 to 14 gauss (a unit of magnetic field strength), extremely powerful by comparison with the 0.3- to 0.8-gauss strength of Earth's magnetic field at the surface. Probably because of the still unknown circulation patterns of the liquid metallic hydrogen in the Jovian interior, Jupiter's magnetic field is far more complex than Earth's. In addition to its dipolar field, the Jovian field, according to Pioneer 11 data, also has quadrupole and octupole movements. That is, components of the main field have four and eight poles, respectively, although these are much weaker than the main dipolar field and have been detected only close to the planet.

Further complications arise from the motion of at least five Jovian satellites that orbit within its magnetosphere. In the course of their orbits, Io, Ganymede, Europa, Callisto, and Amalthea absorb highly charged particles that otherwise might be trapped by Jupiter's magnetosphere. Thus these satellites clear channels through Jupiter's radiation belts. Io sputters material into space that forms a torus of charged particles along its orbit about Jupiter. This torus consists of sulfur, sodium, oxygen, and a few lesser constituents ejected from volcanic activity. Ganymede complicates Jupiter's magnetosphere since that satellite itself possesses a magnetic field and therefore has a magnetosphere within Jupiter's own magnetosphere.

Perhaps because Io in particular possesses an ionosphere that provides it with a conductive fluid, not only

has it trapped charged particles otherwise destined for Jupiter's magnetosphere, but it also produces and accelerates charged particles. When Io is in a fixed position along an Earth-Jupiter line, radio emissions increase; thus, it is believed that its charged particles are an additional source of these emissions. Amalthea has also shown peculiarities as it orbits through the Jovian magnetosphere. Charged particles in its magnetic field unexpectedly do not increase in density toward Jupiter's magnetic equator or to its "surface." Instead, particle density varies widely at many different points.

One of the major reasons that the Galileo spacecraft was outfitted with a sophisticated magnetometer was to attempt to determine how plasma is transported through Jupiter's magnetosphere. Naturally the magnetometer

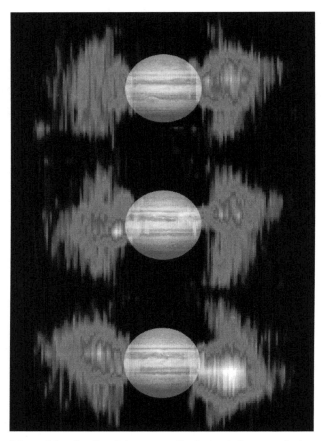

*Mapped by the Cassini spacecraft, Jupiter's inner radiation belts, showing three views during a ten-hour rotation of Jupiter. The image of the planet has been superimposed. Because Jupiter's magnetic field is tilted in relation to the planet's poles, the radiation belts wobble during the planet's rotation. (NASA/ JPL)*

was also intended to provide precise determinations of spatial and temporal variations of Jupiter's magnetic field, and the extent and shape of the planet's complex magnetosphere as well as its interactions with the solar wind.

## Knowledge Gained

Between the recording of Jovian radio emissions in the 1950's and the Pioneer and Voyager robotic space flights more than twenty years later, knowledge of the Jovian magnetic field increased by quantum leaps. New data not only have confirmed or disproved older theories about the field's origins, extent, and inner and outer complexities but also have led to more consistent and plausible theories about the structure of Jupiter itself—indeed, of other planets as well.

Sources and at least some causes of Jupiter's copious radio emissions have been identified. Based on hard, if still incomplete, evidence gathered by American spacecraft, the configurations of the magnetosphere, magnetopause, magnetosheath, and magnetotail have been delineated. The magnetic field's axis and its location have been defined. Quadrupole and octupole fields within the overall dipolar field have been discovered. The strength of the field and fluctuations within it have also been measured. Investigations of the magnetosphere have led to new theories concerning the makeup of the Jovian interior and some of the convective and conductive functions of the liquid metallic hydrogen that composes much of it.

Further, many of the special characteristics of the outer magnetosphere have been explained as the result of two principal mechanisms. First, as observed by James Van Allen (discoverer of Earth's Van Allen radiation belts), a mass of low-energy plasma trapped in the Jovian magnetic field has created pressures that have inflated the field as if it were a balloon. Second, because of interactions with the planet's rotating magnetic field, the plasma corotates (over a period of 9 hours, 55.5 minutes), creating a centrifugal force that contributes to the outward pressure. Plasma analyzers on the two Voyager spacecraft indicated that the plasma originated from gases—principally sulfur dioxide and hydrogen sulfide with sodium and oxygen in lesser amounts—vented by Io's vigorous volcanic activities. This plasma is responsible for the Io torus, a phenomenon unique to Jupiter's magnetosphere. That is, Io is surrounded by a doughnut-shaped band of excited charged particles, some captured from the solar wind as it orbits Jupiter and some produced and agitated by Io's own environment.

It is also understood now that Io and the other inner satellites, in the course of their orbits, attract many particles that enter Jupiter's inner magnetosphere and thereby limit that region's population of trapped particles. Particles trapped in the Jovian magnetosphere also affect the chemistry of the planet's environment. Andrew Ingersoll has shown that trapped particles rain down from the magnetosphere into the Jovian atmosphere. There lightning and other charged particles break down dominant chemical species, thus keeping a balance between production and breakdown of hydrocarbons in the Jovian atmosphere.

The Galileo spacecraft provided long-term measurements of Jupiter's magnetic field and the interaction of the larger satellites with that complex structure. Magnetic measurements combined with gravitational information refined the understanding of Jupiter's interior structure.

## Context

Jupiter attracts scientific attention because it is big, stunningly dynamic in visual wavelengths, and noisy in the radio range. By far the largest body in the solar system after the Sun, it is a natural target of interest for the astronomer's gaze and telescopic observation. Consequently, the planet's general outward appearance—its cloud cover, its bands, and its fascinating Great Red Spot—have been closely observed over several centuries.

An emitter of abundant radio emissions, Jupiter was sufficiently noisy to arouse the curiosity of radio astronomers, who were then led to explore the spectrum of these emissions and theorize about the planet's structure. By the 1950's, thanks to analysis in the visible, infrared, and radio frequency portions of the electromagnetic spectrum, much was known of Jupiter's chemical composition. The light gases hydrogen and helium were dominant, and the sheer size of Jupiter meant that it would require billions of years to dissipate them. Some believed that beneath the planet's dense cloud cover there were mountains, while others believed the planet to be entirely gaseous. In the light of the vast quantities of information gained by the early 1980's, much of the pre-1950 understanding of the planet seems rudimentary, even laughable.

Between the 1950's and the early 1980's, several broad intellectual currents inspired scientific interest in planetary science, in Jupiter, and thereby ultimately in the planet's magnetosphere. One such current was general scientific acceptance of a theory of the solar system's origin that postulates condensation from a disk-shaped nebula of gas and dust. This theory proposes that about 4.6 billion years ago, a cloud of interstellar gas or dust, overwhelmed—as

Carl Sagan and others have suggested—by an exploding star, collapsed and condensed to form the solar system. The central mass in the interstellar formation, contracting under its own gravity, produced such prodigious heat that it generated a thermonuclear reaction, from which the Sun evolved. Lesser masses—Earth, for example—experienced less heating and so, as planets, moons, asteroids, and comets, bathed in and reflected the Sun's light. Substantiation of this theory's accuracy, validity, and reliability required a fresh empirical examination of each major object in the solar system.

With the dawn of the space age, it became possible to make in situ observations or at least the nearest equivalent. Chronologically speaking, data collected by the flybys of Pioneers 10 and 11 and Voyagers 1 and 2, by Galileo in orbit about Jupiter, and by Cassini and New Horizons passing through the Jupiter system, en route to other destinations, vastly increased the storehouse of knowledge concerning Jupiter's magnetosphere. Contributions of these exploratory missions to astronomy and comparative planetology helped advance a greater understanding of Earth's relationship to the rest of the solar family.

*Clifton K. Yearley*

## Further Reading

Bagenal, Fran, Timothy E. Dowling, and William B. McKinnon, eds. *Jupiter: The Planet, Satellites, and Magnetosphere*. Cambridge, England: Cambridge University Press, 2007. A comprehensive work about the biggest planet in the solar system, comprising a series of articles by experts. An excellent repository of photography, diagrams, and figures about the Jupiter system and the various spacecraft missions that unveiled its secrets.

Beatty, J. Kelly, Carolyn Collins Petersen, and Andrew Chaikin, eds. *The New Solar System*. 4th ed. Cambridge, Mass.: Sky, 1999. Filled with color diagrams and photographs, this is a popular work on solar-system astronomy and planetary exploration through the Mars Pathfinder and Galileo missions. Accessible to the astronomy enthusiast, it provokes excitement in the general reader and communicates an appreciation for the need to understand the universe around us.

Cole, Michael D. *Galileo Spacecraft: Mission to Jupiter*. New York: Enslow, 1999. Provides a full description of the Galileo spacecraft, its mission objectives, and science returns through the primary mission. Particularly good at describing mission objectives and goals. Suitable for a younger audience.

Fimmel, Richard O., James Van Allen, and Eric Burgess. *Pioneer: First to Jupiter, Saturn, and Beyond*. NASA SP-446. Washington, D.C.: Government Printing Office, 1980. A classic, informative, and readable account of the first spaceflight into the Jovian environment, offering an excellent synthesis of its amazing findings. Includes illustrations, tables, a modest bibliography, and an index.

Harland, David H. *Jupiter Odyssey: The Story of NASA's Galileo Mission*. New York: Springer Praxis, 2000. This book provides, in a single work, virtually all of NASA's press releases and science updates during the first five years of the Galileo mission. Contains an enormous number of diagrams, tables, lists, and photographs. Provides a preview of the Cassini mission (which did not end until after publication).

Hartmann, William K. *Moons and Planets*. 5th ed. Belmont, Calif.: Thomson Brooks/Cole, 2005. An updated version of a classic text that covers all aspects of planetary science. Jupiter is discussed in comparison to the other planets, as the text takes a comparative planetology approach rather than providing individual chapters on each planet in the solar system.

Ingersoll, Andrew P. "Jupiter and Saturn." In *The Planets*, edited by Bruce Murray. San Francisco: W. H. Freeman, 1983. An authoritative discussion that is profitable for both specialists and educated laypersons. Includes superb photographs and graphics. At the end of the book, there are biographies of Ingersoll and other authors, as well as a select bibliography for this and other chapters. The volume covers only what was being learned from the Pioneer and Voyager spacecraft on their flybys of Jupiter, but it is an excellent source for early understandings of the two gas giants.

Irwin, Patrick G. J. *Giant Planets of Our Solar System: An Introduction*. 2d ed. New York: Springer, 2006. Suitable as a textbook for upper-level college courses in planetary science. Focuses on Jupiter, Saturn, Uranus, and Neptune and their satellites, rings, and magnetic fields. Filled with figures and photographs. Accessible to the serious general audience.

McBride, Neil, and Iain Gilmour, eds. *An Introduction to the Solar System*. Cambridge, England: Cambridge University Press, 2004. A complete description of solar-system astronomy suitable for an introductory college course but accessible to nonspecialists as well. Filled with supplemental learning aids and solved student exercises. A companion Web site is available for educator support.

Stone, Edward C., and A. L. Lane. "Voyager 2 Encounters with the Jovian System." *Science* 206 (November 23, 1979): 925-927. Technical, but readily understandable by readers with a basic background in mathematics and physics. Discusses data generated about the Jovian magnetosphere and other features of the planet (and its moons). Stone was chief scientist for the Voyager mission. Contains illustrations, tables, and a brief bibliography.

# JUPITER'S RING SYSTEM

**Categories:** Planets and Planetology; The Jovian System

*Jupiter's ring system consists of four relatively dull, ethereal rings composed of submicron- to micron-sized dust grains. This system provides important clues and insights into the processes that are involved in the generation of circumstellar disks around planets. In the case of Jupiter, the primary mechanism that produces and replenishes its rings is dust generated when interplanetary meteoroids collide with four of Jupiter's small, inner satellites.*

## Overview

Trailblazing missions to explore Jupiter and Saturn were conducted by Pioneer 10 and 11 and Voyager 1 and 2. When Pioneer 11 flew by Jupiter in 1973-1974, observations of rather rapid variations in the number of charged particles orbiting Jupiter at specified distances from the planet suggested the possibility of a ring system that might be absorbing the particles. Although the Pioneer spacecraft were not sufficiently stabilized to facilitate taking images, Voyager 1 and 2 were. On March 4, 1979, an overexposed image from Voyager 1 finally confirmed the existence of a ring system around Jupiter, a result long anticipated by astronomers. Voyager 2 cameras captured numerous pictures of Jupiter's ring system at geometries and resolutions that were previously unobtainable. Three separate rings were discovered, the central Main Ring, the inner Halo Ring, and the outer Gossamer Ring. These rings exist within the Roche limit, the distance from the planet to where tidal forces prevent ring particles from forming into aggregates due to gravitational attraction.

On October 18, 1989, the 2.7-ton Galileo spacecraft, consisting of the main body orbiter and a probe, was launched. On December 7, 1995, the orbiter reached Jupiter and made thirty-four orbits around the planet before plunging into the planet's atmosphere in 2003. Using a solid-state imaging camera, high-quality images were taken of Jupiter's satellite-ring system. After careful analysis of the pictures, it was concluded that Jupiter's ring system is formed from dust generated as high-speed interplanetary micrometeoroids collide with the planet's four small inner satellites (sometimes called moonlets)—Metis, Adrastea, Amalthea, and Thebe—which orbit within the rings. A totally unexpected result was that the outermost Gossamer Ring consisted of two rings, one embedded inside the other. The outermost Gossamer Ring was bounded by the orbit of Thebe and named the Thebe Gossamer Ring, while the innermost one was bounded by the orbit of Amalthea and named the Amalthea Gossamer Ring.

The Cassini spacecraft, designed to obtain high-resolution images of planetary ring systems, made its closest approach to Jupiter on December 30, 2000. It imaged Jupiter's rings using different wavelengths that provided further constraints on the size, distribution, shapes, and composition of the particles within the rings. The reddish

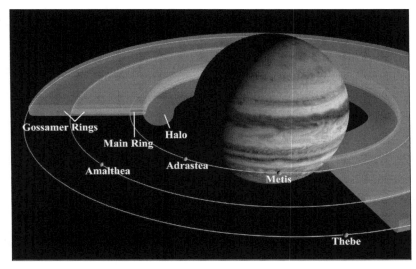

*The diagram displays the main components of Jupiter's ring system as well as the small inner moons, from which the dust of the rings originates.* (NASA/JPL/Cornell University)

colors of the Jovian ring particles indicated a silicate or carbonaceous composition, just like that of the small embedded satellites. Images showed that the particles in Jupiter's rings were nonspherical. Cassini images also captured the motion of the two Gossamer Ring satellites, Thebe and Amalthea.

Images of Jupiter's rings taken by the New Horizons spacecraft in early 2007 confirmed that the dusty Jovian ring system was being replenished continually from embedded source bodies. The Main Ring was found to consist of three ringlets, one just outside the orbit of Adrastea, one just inside the orbit of Adrastea, and one just outside the orbit of Metis. Boulder-sized clumps, consisting of a close-paired clump and a cluster of three to five clumps, were discovered in the Main Ring just inside the orbit of Adrastea. Although the origin and nonrandom distribution of the clumps remain unexplained, they are confined to a narrow belt of motion by the gravitational influence of the two innermost satellites of Jupiter. New Horizons images established a lower limit to the diameter of Jupiter's moons of 0.5 kilometer.

From observations, measurements, and numerical modeling methods applied to data collected from Voyager 2, Galileo, Cassini, New Horizons, the Hubble Space Telescope (HST), and the Keck telescope, it has been concluded that Jupiter's rings are extremely tenuous and contain significant amounts of short-lived dust. In addition to the gravitational perturbations produced by the small satellites embedded within and bounding Jupiter's ring system, the dynamics of its faint, ethereal dusty rings are dominated by effects that involve electromagnetic forces, solar radiation pressure, and various drag forces. In the process of conserving angular momentum, the rapid spin rates tend to flatten the rings. The relatively bright, narrow Main Ring has a rather sharp outer boundary that coincides with the orbit of Adrastea. Just inside this boundary is the orbit of Metis. Since the Main Ring extends only inward from these small source moonlets, it has been concluded that particles in the Main Ring must drift inward. The width of the Main Ring is approximately 6,440 kilometers, with a thickness that varies between 30 and 300 kilometers. Dust size ranges from 0.5 to 2.5 microns in diameter.

Interior to the Main Ring lies a thick torus of particles known as the Halo Ring. Its thickness is primarily determined by Jupiter's very strong magnetic field operating on the ring's submicron dusty grains. The thickness of the Halo Ring is approximately 12,500 kilometers, while its width is about 30,500 kilometers. In visible light, the Halo Ring has a bluish color. The very faint Almathea Gossamer Ring has an estimated width of 53,000 kilometers and a thickness of 2,500 kilometers. It has been imaged from the Earth using the Keck telescope. It appears brighter near its top and bottom edges and also brightens toward Jupiter. The dust grain size in this ring is similar to that in the Main Ring. The faintest Jovian ring, the Thebe Gossamer Ring, has a width of approximately 97,000 kilometers and a thickness of 8,500 kilometers. Dust grain size varies from 0.2 to 3.0 microns. The Thebe Gossamer Ring is observed to extend beyond Thebe, which is apparently due to coupled oscillations produced by time-varying electromagnetic forces that cause the ring to extend outward. The thickness of each Jovian ring is primarily controlled by the inclination of the orbit of its embedded moonlet.

## Knowledge Gained

The existence of Jupiter's ring system was unambiguously determined in March, 1979, by Voyager 1. Until that time, most astronomers and astrophysicists were confounded as

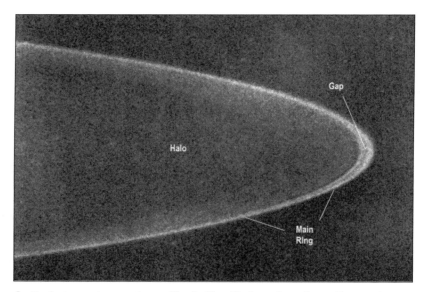

*Jupiter's ring system was imaged by the New Horizons Long Range Reconnaissance Imager on February 24, 2007.* (NASA/Johns Hopkins University Applied Physics Laboratory/Southwest Research Institute)

to why Saturn had a ring system but Jupiter did not. In July, 1979, more detailed images from Voyager 2 showed that the dull, diffuse ring system of Jupiter consisted of three separate rings. The ring system exists within an intense radiation belt of electrons and ions that are trapped in Jupiter's magnetic field. Resulting drag forces play an important role in determining the motion of the ring particles.

Images obtained from the Galileo spacecraft between 1995 and 2003 provided increasing detail about Jupiter's rings. The shape, width, thickness, optical depth, and brightness of each ring were determined, as well as dust spatial densities, grain sizes, and grain collision speeds. Jupiter's faint, dark, narrow rings (albedo about 0.05) consist of submicron- to micron-sized rock fragments and dust but do not contain ice, as do Saturn's rings. The number of separate rings in Jupiter's ring system was found to be four when it was determined that the Gossamer Ring consisted of two distinct rings. Further constraints on the composition, distribution, size, and shape of particles within Jupiter's rings were established in 2000 by the Cassini probe. In 2007, images from the New Horizons spacecraft revealed the fine structure of Jupiter's Main Ring. It consists of three ringlets and contains two families of boulder-sized clumps.

From the variety of measurements, observations, and analysis of collected data, Jupiter's ring system is now the best understood prototype of planetary ring systems that consist of thin, diffuse sheets of dusty debris that are primarily generated by small source moonlets. The relative motion of the dust grains within Jupiter's rings and the orientation of the orbits of the rings are primarily controlled by three processes: the spinning, asymmetric, very strong magnetic field of Jupiter; the absorption, reemission, and scattering of solar radiation energy by the dust particles, which produces momentum changes that induce orbital changes; and drag forces on the grains produced by solar radiated photons, as well as by ions and atoms that are orbiting around Jupiter. Since dust particles are continually being removed from Jupiter's rings by these processes and then replenished by dust from the four inner satellites, the dust grains existing in the rings are estimated to be relatively young, probably much less than one million years old. As dust particles are ejected from the moonlets, the particles enter orbits like those of the moonlets, which causes the rings to wobble up and down as they orbit around Jupiter's equator. Micrometeoritic impacts that generate dust from the moonlets also color, chip, erode, and fragment the dust particles within the rings.

## Context

Spacecraft flybys and orbiters of Jupiter and Saturn have greatly increased the scientific understanding of planetary rings. Numerical methods have been employed to simulate the physical processes occurring within Jupiter's rings by including collisional, gravitational, and electromagnetic interactions among the orbiting ring particles. Resulting models are providing keys to help guide observational strategies for future space missions.

The ring system of Jupiter provides insights into the characteristics of flattened systems of gas and colliding dust particles that are analogous to those that have eventually resulted in the formation of solar systems. In particular, Jupiter's rings offer an accessible laboratory for observing, measuring, and modeling the ongoing processes similar to those associated with the circumstellar disks that were most likely active in the solar nebula disk when the solar system containing the Earth was formed.

Further analysis, detailed examination, and numerical modeling of the data acquired by the Cassini probe and New Horizons spacecraft should provide more high-resolution maps, identify the detailed radial structure of Jupiter's ring system, and reveal invaluable time-variable features associated with the evolution of the rings. Future observations and measurements will offer new insights into the dynamic forces that shape and maintain these fascinating structures. The National Aeronautics and Space Administration (NASA) is considering a mission to Jupiter to explore the planet and its satellite-ring system in detail from a polar orbit. NASA also plans to develop small spacecraft with the capability of hovering over the rings of Jupiter and Saturn, which should provide the necessary additional data and insights for producing refined models and an advanced understanding of planetary ring structures and why they vary vastly among the gas giants.

*Alvin K. Benson*

## Further Reading

Bagenal, Fran, Timothy E. Dowling, and William B. McKinnon, eds. *Jupiter: The Planet, Satellites, and Magnetosphere.* Cambridge, England: Cambridge University Press, 2004. The physical characteristics, temperature, atmospheric makeup, and satellite and ring systems of Jupiter are clearly analyzed. A description of spacecraft missions to Jupiter, numerous high-quality, full-color photographs, and many figures, tables, and diagrams elucidate a tour of the Jovian system.

Elkins-Tanton, Linda T. *Jupiter and Saturn*. New York: Chelsea House, 2006. Discusses the role that Jupiter has played in advancing our understanding of the planets that orbit the Sun. Clearly describes the discovery of Jupiter's ring system, details the Jupiter satellite-ring system, explains why the rings exist, and contrasts the ring systems of Jupiter and Saturn. An appendix lists all the known satellites of Jupiter and Saturn.

Esposito, Larry. *Planetary Rings*. Cambridge, England: Cambridge University Press, 2006. This treatise covers all aspects of planetary ring systems, including ring history, physical processes involved in ring evolution, and mathematical models used to describe them. In particular, Esposito discusses Jupiter's ring-satellite system, the age of Jupiter's rings, and the size distribution of Jupiter's rings. The text is clearly written, illustrated with many diagrams and images, and geared for a wide reading audience.

Harland, David M. *Jupiter Odyssey: The Story of NASA's Galileo Mission*. Chichester, England: Springer Praxis, 2000. A detailed account of the long trek to Jupiter by Galileo and its five years of exploration within the Jovian system. Spectacular results are presented from the observations and measurements of the satellite-ring system of Jupiter. Written for general readers, undergraduates, and faculty alike, the book is very well illustrated and includes references to many relevant Web sites.

Krüger, Harald. *Jupiter's Dust Disc: An Astrophysical Laboratory*. Aachen, Germany: Shaker-Verlag, 2003. On the basis of spacecraft observations and measurements made by Galileo, Krüger delineates the physical processes and mechanisms involved in producing Jupiter's ring system. The data clearly indicate that the Gossamer Ring material comes from Jupiter's small moonlets Thebe and Amalthea.

Miner, Ellis D., Randii R. Wessen, and Jeffrey N. Cuzzi. *Planetary Ring Systems*. New York: Springer Praxis, 2006. Provides comprehensive coverage of the scientific significance of ring systems, ring characterization and comparison, and the history of the discovery of planetary ring systems, including observations of Jupiter's ring system from Voyager 1, Voyager 2, and Galileo. Various theories for the formation of planetary ring systems are explored.

# JUPITER'S SATELLITES

**Categories:** Natural Planetary Satellites; Planets and Planetology; The Jovian System

*Jupiter has many natural satellites, among which are some of the largest and most intriguing of the solar system. Pioneer 10, Pioneer 11, Voyager 1, Voyager 2, and Galileo collected photographs and other vital information about several of these Jovian satellites, which have been used to describe the histories of these bodies and the solar system in general.*

### Overview

The four largest satellites of Jupiter, Io, Callisto, Europa, and Ganymede, were discovered in 1610 by Galileo Galilei and have been objects of curiosity for astronomers since that time. By 1990 astronomers had discovered sixteen moons around Jupiter—fourteen using Earth-based telescopes and two using photographs from Voyager. One of the fourteen was confirmed by the Voyager mission. After the arrival of the Galileo spacecraft in orbit about Jupiter, and in cooperation with space-based and Earth-based telescope observations, the discovery of Jovian satellites increased regularly. As of 2008, Jupiter had sixty-three recognized satellites. Satellites of the outer planets are classified as "regular" if their orbits are nearly circular and are in, or near, the plane of the equator of the planet. "Irregular" satellites have highly elliptical orbits, have the orbital plane tilted relative to the equator, or revolve in a retrograde (westward) direction.

Jupiter's satellites can be divided into three groups. The largest satellites, the often-called Galilean moons, are all regular satellites, as are the four small satellites whose orbits are nearer to Jupiter than is Io's. Eight irregular satellites lie beyond Callisto, and four of them have retrograde revolutions. Only the Galilean satellites and Amalthea have been examined closely by spacecraft. The other satellites are too small and too distant from the probes that have passed through the Jovian system to be effectively studied.

Many of the outer planets are known to have icy surfaces rather than the rocky surfaces that charaacterize the terrestrial planets (Mercury, Venus, Earth, and Mars) and their satellites. Ice behaves somewhat like a plastic unless its temperature is less than 133 kelvins—that is, it will flow and not retain a structure over long periods of time. This process is called plastic deformation. Rocky surfaces do not undergo plastic deformation. It is expected,

*A composite image showing the "Galilean" satellites of Jupiter—the four largest (left to right): Ganymede, Callisto, Io, and Europa. The Callisto image is from Voyager's 1979 flyby of Jupiter; the rest were taken in 1996 by the Galileo spacecraft.* (NASA/JPL/DLR)

therefore, that the terrestrial planets and satellites would retain structural features such as craters indefinitely, while an icy satellite might not be able to do the same unless the temperature is always below 133 kelvins.

Amalthea is the innermost of Jupiter's sizable satellites. It is small and elongated, with a length of 270 kilometers and an average diameter of 155 kilometers. Its long axis is pointed toward Jupiter throughout its orbit. The composition of its dark red surface and of its bulk is unknown. Four features were found by Voyager—two craters, Pan and Gaea, and two mountains, Mons Ida and Mons Lyctas. Nothing definite is known about these features. The scarred surface reflects a history of bombardment.

Io is the Galilean satellite nearest Jupiter. It is a rocky satellite with little or no remaining water. In fact, it may be the driest place in the outer solar system. Its density is similar to that of Earth's moon, so it is assumed that its bulk composition must also resemble that of the Moon. Its surface features, however, are dramatically different from those of the Moon. The surface displays materials that are yellow, red, brown, black, and white. Io shows no cratering but contains layered lava plains, volcanic mountains, calderas, and other evidence of volcanism. Of the nine active volcanoes found by Voyager 1, seven were still active when Voyager 2 passed. The highest mountains were about 9 kilometers tall, making them the largest found on the Jovian satellites. The polar areas are darker than the remainder of Io and show evidence of long periods of deposition, faulting, and erosion. Significant changes in Io's volcanic activity were noted

when the Galileo spacecraft arrived in 1996, fifteen years after Voyager 1 passed by.

Io's volcanic activity takes on several forms. Plume activity is associated with the region 45° north and south of the equator. Material in the plumes is ejected at a velocity of 3,200 kilometers per hour. Much of this material is sulfur dioxide, which then crystallizes and falls back to the surface as a white solid. The material in the plumes is apparently carried to the surface along fissures in the surface rather than through pipe vents like those characteristic of volcanoes found on Earth. Some of Io's volcanoes produce a lava flow rather than a plume. Hot spots exist on the surface where a large amount of energy is being transferred from the interior of Io but where the quantity of volatile material available is insufficient to produce a plume. The location of such hot spots is not limited to the band in which the plume volcanoes are found. As a consequence of the extensive volcanic activity, the surface of Io changes rapidly. Significant details changed between the flyby of Voyager 1, the flyby of Voyager 2, and the arrival of the orbiting Galileo spacecraft. Galileo discovered that Io has an iron core.

Europa also provided surprises for the Voyager scientists. Density measurements imply that the moon is about 90 percent rocky core and 10 percent water ice as the crust. Because of its nearness to Jupiter, it was expected to have an extensively cratered surface. Voyager photographs of Europa, however, show one of the smoothest surfaces in the solar system. Very few craters are visible. The surface is marked with long, narrow lines that resemble the cracking pattern of an egg, along with dark spots and mottling. Lines are regions where the crust cracked and water from the interior was squeezed out onto the surface. A reflectivity of 64 percent is an indicator of quite pure water ice on the surface of Europa. Since an old surface would be expected to collect dust and other space debris and produce a dirty surface with low reflectivity, the pure water ice implies that there is some process taking place that continually renews Europa's surface. After Galileo's close flybys in 1996 and 1997, scientists concluded the ice crust is no thicker than 150 kilometers and floats on a liquid ocean.

Ganymede is the largest of the Jovian satellites, although it is not much larger than Callisto. Ganymede has a metallic core surrounded by a layer of ice and silicates, and its crust is probably thick water ice. The metallic core generates a magnetic field. About one-half of Ganymede's surface has a dark, cratered terrain, while the remaining half is much lighter and has fewer craters.

Lighter areas of Ganymede have a series of parallel mountains and valleys reminiscent of the low Appalachian Mountains in North America. These may have developed from long cracks or faults that are separated by strips of land that have alternately been lifted or depressed. Depressed areas could have flooded with liquid water from the interior. This type of mountain building results from tension in the crust rather than the compression of the crust that is believed to have caused the Appalachian Mountains to be uplifted. Reflectivity of the lighter areas is about 40 percent, while in the more heavily cratered areas it is only 25 percent. Bright ray craters are found in all parts of the moon. (A ray crater is one in which the debris cast out of the crater at the time

of impact is distributed along radial directions like rays or beams.) These craters are evidently the result of water from under the crust that has splashed out on the surface when some object collided with Ganymede. A dome 260 kilometers in diameter and 2.5 kilometers in height which is surrounded by a number of small craters may be evidence of water volcanism. Impacts that formed the small craters may have weakened the crust, so that water flowed up through newly formed cracks and holes. There are very few large craters, but some ghost craters exist. These show the details of the crater, but actual physical features such as walls have disappeared because of plastic deformation of Ganymede's surface.

Callisto is the outermost of the Galilean satellites and probably the least active. Its surface temperature ranges from 150 kelvins at noon to 100 kelvins before dawn. At temperatures as cold as these, a layer of ice only 1 meter thick would take 4.5 billion years to evaporate. The dark, heavily cratered surface has a reflectivity of only 18 percent, which implies that the surface ice has a low purity. Callisto's landscape is almost exclusively the product of impacts. The crater features are subdued, and no large impact basins exist. This relative smoothness is probably the result of the plastic deformation of the crust. In the Valhalla basin, visible rings on the surface may be the remnant of a very large impact basin. Few ray craters are visible.

**Knowledge Gained**

The composition of the surface of Io has been reasonably well established. The white solid is sulfur dioxide condensed from plume activity. The remainder of the surface

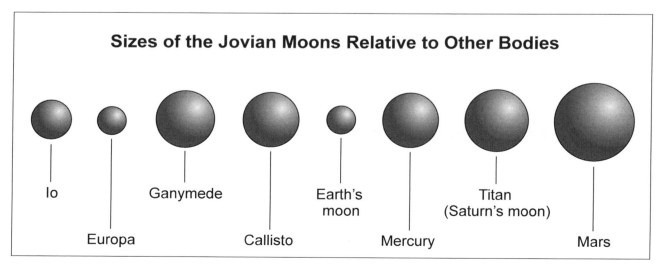

**Sizes of the Jovian Moons Relative to Other Bodies**

Io  Europa  Ganymede  Callisto  Earth's moon  Mercury  Titan (Saturn's moon)  Mars

## Some Facts About the Major Jovian Satellites

| | Mass (1020 kg) | Radius (km) | Mean Density (kg/m³) | Orbital Period (days) |
|---|---|---|---|---|
| **Jupiter's Major Moons** | | | | |
| Io | 893.3 | 1,821.3 | 3,530 | 1.77 |
| Europa | 479.7 | 1,565 | 2,990 | 3.55 |
| Ganymede | 1,482 | 2,634 | 1,940 | 7.15 |
| Callisto | 1,076 | 2,403 | 1,851 | 16.69 |
| **Saturn's Major Moons** | | | | |
| Mimas | 0.375 | 209 | 1,140 | 0.94 |
| Enceladus | 0.73 | 256 × 247 × 245 | 1,120 | 1.37 |
| Tethys | 6.22 | 536 × 528 × 526 | 1,000 | 1.89 |
| Dione | 11.0 | 560 | 1,440 | 2.74 |
| Rhea | 23.1 | 764 | 1,240 | 4.52 |
| Titan | 1,345.5 | 2,575 | 1,881 | 15.94 |
| Hyperion | 0.2 | 185 × 140 × 113 | — | 21.28 |
| Iapetus | 15.9 | 718 | 1,020 | 79.33 |
| **Uranus's Major Moons** | | | | |
| Miranda | 0.66 | 240 × 234.2 × 232.9 | 1,200 | 1.41 |
| Ariel | 13.4 | 581.1 × 577.9 × 577.7 | 1,670 | 2.52 |
| Umbriel | 11.7 | 584.7 | 1,400 | 4.14 |
| Titania | 35.2 | 788.9 | 1,710 | 8.70 |
| Oberon | 30.1 | 761.4 | 1,630 | 13.46 |
| **Neptune's Major Moons** | | | | |
| Naiad | — | 29 | — | 0.29 |
| Thalassa | — | 40 | — | 0.31 |
| Despina | — | 74 | — | 0.33 |
| Galatea | — | 79 | — | 0.43 |
| Larissa | — | 104 × 89 | — | 0.55 |
| Proteus | — | 218 × 208 × 201 | — | 1.12 |
| Triton | 214.7 | 1,352.6 | 2,050 | −5.88 |
| Nereid | 0.2 | 170 | 1,000 | 360.14 |

*Notes:* Moons are listed in order of their distance from the planet. A minus sign preceding orbital period signifies retrograde motion.

*Source:* Data are from the National Aeronautics and Space Administration/Goddard Space Flight Center, National Space Science Data Center.

is largely elemental sulfur forms, which can have a variety of colors, depending upon the process occurring during solidification. Sodium has been detected in the space around the satellite. The existence of active volcanoes on Io requires the presence of molten material under the surface. Tidal action of Jupiter and of other Jovian satellites upon Io is probably the main source of heat energy that generates molten material. Although Jupiter exerts a very large attractive force on Io, the gravitational attractions of the other satellites cause the surface of Io to be pulled in competing directions. Friction resulting from the flexing of the surface causes the material beneath the crust to melt. The thickness of the crust is not precisely known, but it is thought to be minimal. A sulfur sea may exist below the crust. An additional source of heat is electric current induced in the iron sulfide core of the satellite as it interacts with the strong magnetic field of Jupiter. A portion of the radio waves emitted by the Jovian system originate from Io's interaction with the magnetic field of Jupiter and with the

plasma disk (a disk of charged particles) that surrounds the planet.

Europa is another Jovian satellite that shows signs of activity, although how recent that activity may have been is not known. Some scientists believe that there may be an ocean of liquid water as much as 100 kilometers deep lying below the icy crust of Europa. Occasional cracks in the crust allow the water to flow onto the surface, erasing any evidence of past impacts, when large flows occur, or simply to come up through the cracks to form the ridges that mark the surface.

Ganymede also shows evidence of having been active after the major cratering epochs ended within the solar system. Mountains formed from tension in the surface are

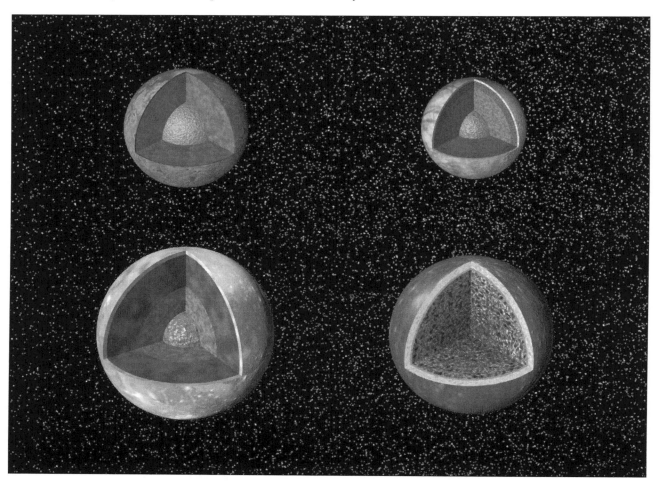

*An artist's cutaway views of the interior structures of the four major Jovian satellites, as implied by gravitational and magnetic field experiments (clockwise from upper left): Io, Europa, Callisto, and the largest, Ganymede. Callisto is the only one of the four that does not show evidence of a metallic iron-nickel core surrounded by a mantle of rock. Ganymede and Europa are believed to have outer shells of water (ice or liquid). Callisto is believed to consist of a mixture of ice and rock.* (NASA/JPL)

evidently a consequence of a series of internal upheavals during the earlier life of the satellite. It is possible that these changes were initiated by changes in the crystal structure of the ice as the core of Ganymede slowly cooled. Resulting expansions and contractions could have cracked the surface and allowed the lower areas to be flooded with a lavalike flow of liquid water.

Callisto is believed to have been inactive since the initial formation of its crust. It is generally accepted that Ganymede and Callisto are differentiated objects. This means that initially the objects were a uniform mix of water ice and rocky materials. This mixture was heated from some source, perhaps radioactivity of the rocky material, and the denser, rocky material settled to the core, leaving the less dense water on the surface, where it eventually froze and formed the crust. For Callisto, it seems that this development brought an end to its self-generated geologic activity. All Callisto's other topographic features resulted from collisions with other space objects.

## Context

Although the Galilean moons have been known since 1610, because of their small size astronomers initially found it difficult to identify details on the surface of any of them. Photographs taken by equipment carried high into Earth's atmosphere showed Io to have dark polar regions. Careful measurements of brightnesses indicated that there was a variation as the position in orbit changed, suggesting that the surface on a given satellite did not have a uniform reflectivity. These data also led astronomers to the conclusion that the satellites were tidally locked to Jupiter, so that their periods of rotation were equal to their periods of revolution.

Another interesting feature of the orbits is that the periods of revolution of Io, Europa, and Ganymede are tied together by gravitational coupling. The period for Europa is twice that for Io, and the period for Ganymede is twice that for Europa. If any one of these three orbits changed, the other two would also change to restore the ratios that currently exist. One difficulty in the models that describe the origin of Jupiter is that they offer no account of a process through which these three satellites could move into such an orbital relationship.

Voyager data revealed four satellites that were different from expectations in many ways and that had significant differences among themselves. Models have been constructed that attempt to describe processes that would lead to the formation of the moons and give rise to their current conditions. The region in which Io and Europa are believed to have been formed would have had a fairly high temperature, and volatile materials such as water would not have been present in significant quantities. The tidal heating that Io experiences has driven off any volatile materials, such as water and carbon dioxide, that may have been a part of the original body.

Europa retained more water at formation than did Io and has not lost it. There appears to be enough heat in the silicate core to keep much of the water in the liquid phase and to allow it to flow through cracks that form in the moon's surface. This flow process has essentially resurfaced Europa during its history.

Ganymede has a mass that is roughly 50 percent water. Much of this water is in liquid form. It is assumed that the core is warm enough to transfer energy to the water to keep it liquid. The water's liquid state allows parts of the icy surface to be renewed.

Callisto's icy surface is covered with significant quantities of silicate materials. Although there may be a layer of liquid water in the interior, the crust is thick enough to prevent any of the liquid from reaching the surface. The process of plastic deformation has eliminated the features of the larger impact basins, although some of the materials that would have been part of the basin are visible in the icy crust, giving the impression that the moon has a large bull's-eye drawn on it.

The next stage in the investigation of Jupiter's satellites would be to send to Jupiter a spacecraft capable of performing detailed investigations over prolonged periods in close orbital proximity to the satellites rather than just doing multiple flybys. Such a mission was proposed and initially funded. The Jupiter Icy Moons Orbiter (JIMO) would have used nuclear propulsion not only to enter the gravitational sphere of Jupiter but also to go from orbit around one satellite to another over a very prolonged mission lasting perhaps a decade. When the program met technical difficulties and costs threatened to rise well above the allocated $1 billion, JIMO was canceled. Nevertheless, among planetary scientists, an orbiter around Europa and perhaps other Galilean moons remains a high priority for understanding the Jovian system.

*Dennis R. Flentge*

## Further Reading

Bagenal, Fran, Timothy E. Dowling, and William B. McKinnon, eds. *Jupiter: The Planet, Satellites, and Magnetosphere.* Cambridge, England: Cambridge University Press, 2007. A comprehensive work about the biggest planet in the solar system. A series of articles provided by recognized experts in their field of study. Excellent repository of photography, diagrams, and figures about the Jupiter system and the various spacecraft missions that unveiled its secrets.

Bredeson, Carmen. *NASA Planetary Spacecraft: Galileo, Magellan, Pathfinder, and Voyager*. New York: Enslow, 2000. This book is part of Enslow's Countdown to Space series. Provides an overview of NASA planetary exploration during the last two decades of the twentieth century. Suitable for all audiences.

Cole, Michael D. *Galileo Spacecraft: Mission to Jupiter*. New York: Enslow, 1999. Provides a full description of the Galileo spacecraft, its mission objectives, and science returns through the primary mission. Particularly good at describing mission objectives and goals. Suitable for a younger audience.

Davies, Ashley Gerard. *Volcanism on Io: A Companion with Earth*. Cambridge, England: Cambridge University Press, 2007. Provides the full analysis of Io's diverse set of active volcanoes based on all spacecraft data and remote sensing from Earth. Technical but accessible to the general science reader too.

Fischer, Daniel. *Mission Jupiter: The Spectacular Journey of the Galileo Spacecraft*. New York: Copernicus Books, 2001. Suitable for a wide range of audiences. Thoroughly explains all aspects of the science and engineering of the Galileo spacecraft. Particularly good are the discussions about the nature of the Galilean satellites.

Geissler, Paul E. "Volcanic Activity on Io During the Galileo Era." *Annual Review of Earth and Planetary Sciences* 31 (May, 2003): 175-211. The definitive work describing the physics and planetary geology of volcanoes on Io. Provides a complete picture of Voyager and Galileo spacecraft results.

Greenberg, Richard. *Europa the Ocean Moon: Search for an Alien Biosphere*. Berlin: Springer, 2005. A complete description of knowledge of Europa through the post-Galileo spacecraft era. Discusses the astrobiological implications of an ocean underneath Europa's icy crust. Well illustrated and readable by astronomy enthusiasts and college students.

Harland, David H. *Jupiter Odyssey: The Story of NASA's Galileo Mission*. New York: Springer Praxis, 2000. Provides virtually all of NASA's press releases and science updates during the first five years of the Galileo mission in a single work. Provides a preview of the Cassini mission. An enormous number of diagrams, tables, lists, and photographs.

Leutwyler, Kristin, and John R. Casani. *The Moons of Jupiter*. New York: W. W. Norton, 2003. Casani was the original Galileo program manager. This volume offers a heavily illustrated discussion of the Galilean satellites and a number of the less well known Jovian moons. The authors attempt an artful text accompanying the scientific findings, which may or may not be to the taste of all readers.

Lopes, Rosaly M. C., and John R. Spencer. *Io After Galileo: A New View of Jupiter's Volcanic Moon*. Heidelberg: Springer, 2007. Another member of the Springer Praxis Space Exploration series, this book summarizes the knowledge gained by Galileo about Io. Suggests new investigations needed to explain those questions that remain about the volcanism of this Jovian moon. Technical.

# MIRANDA

**Categories:** Natural Planetary Satellites; The Uranian System

*Miranda is the one satellite of Uranus about which there is good information. The whole Uranus system is different because Uranus is tilted. Miranda is different from any of the other natural satellites that have been studied.*

### Overview

Miranda is the eleventh of the twenty-seven known moons of Uranus, a "gas giant" planet that is the seventh planet from the Sun. Miranda was discovered by a Dutch-born American astronomer, Gerald Peter Kuiper, in 1948. Miranda was the daughter of the magician Prospero in William Shakespeare's play *The Tempest*. Features and places on Miranda are named for characters in Shakespeare's plays. When Miranda was discovered, photographic emulsions and techniques had just developed to the stage to be very useful in astronomy. Kuiper was using the McDonald Observatory to photograph the four satellites of Uranus that were known at that time: Oberon, Titania, Umbriel, and Ariel. Upon developing the film, he found a bright spot. A few days later, it was proven that the spot could not be a star and determined that it was, rather, a satellite of Uranus.

Miranda is synchronous with Uranus; that is, Miranda presents the same face to Uranus all the time. As a synchronous satellite, Miranda revolves about Uranus in its orbit in the same time that it rotates about its own axis: 33.9 hours (1.41 days). Miranda turns in the same direction as Uranus; that is, it has a prograde rotation. Miranda's orbit is about four degrees out of the plane of Uranus's equator; that is, its inclination is 4.3°. The inclination is large enough to cause doubt about whether Miranda was formed by the same process by which Uranus was formed, or if Miranda instead was captured in a close encounter.

Because Uranus is tipped on its side, Miranda's orbit is almost perpendicular to the orbit of Uranus. Miranda's eccentricity is small, meaning that the orbit is close to circular. The larger the eccentricity, the more elliptical an orbit; an eccentricity of zero means a circular orbit. Miranda's orbit is only slightly elliptical, with the maximum distance from Uranus being 132,000 kilometers and the minimum distance 128,000 kilometers. Miranda is a triaxial ellipsoid, with a 480-kilometer diameter along the axis pointed at Uranus. The equatorial axis is 468 kilometers, and the pole-to-pole axis is 466 kilometers.

The surface of Miranda is so fractured that it seems as if Miranda was torn apart and the pieces put back together in a haphazard manner. Some scientists have stated that Miranda was torn apart and reformed at least five times. Other ideas are that tidal forces caused partial differentiation. Differentiation is a separation of materials, with heavier rock sinking and forcing water ice to the surface. Near-infrared spectrometry has confirmed that the surface is ice. The surface shows it to be intensely cratered from meteor bombardment, although some of the craters have been reduced in height either by a change in the surface or by material ejected by other meteors covering up the crater as the material fell back to the surface. One crater is 25 kilometers across. There are more craters on Miranda than on other outer moons of Uranus. That is to be expected, since the closer a satellite is to the planet and its gravitational pull, the greater the density of meteors it encounters.

There are also coronae—large, 400-kilometer-wide areas of alternating light and dark stripes that are unlike anything seen in the universe. They look like racetracks with several tracks side by side. There are three of these on Miranda, named Arden, Inverness, and Elsinore. From space, Inverness looks like a chevron of bright material on a dark surface. The edges of the Arden and Inverness coronae are a trench with a cliff surrounding the coronae. These cliffs can be 10 to 20 kilometers high. The coronae seem to be at a lower altitude than the surrounding surface.

Two of the coronae, Arden and Inverness, have albedos (reflectivity) that differ from that of the surrounding material. Elsinore does not have the exterior trench and has much the same albedo as the surrounding material. The coronae were formed in order: first Arden, then Inverness, and then Elsinore. Arden is in the leading hemisphere. Elsinore is in the trailing hemisphere. The sharp tip of the chevron edge of the Inverness corona is very close to the south pole of Miranda.

There are fracture lines and gorges running across the older cratered surface. There are few craters in the coronae, suggesting that they are not as old as the cratered surface area. Some scientists draw a correlation between the partial differentiation and the coronae. The coronae are a feature of partial differentiation, caused by a Miranda-Umbriel-Ariel orbital resonance. For some reason, the differentiation did not finish, but the rise of ice and sinking of heavier material formed the coronae.

Two other satellites of Uranus, Umbriel and Ariel, seem to interact with Miranda. A gravitational pull occurs when the satellites are close and then dissipates when they are far apart. This pull, along with the gravitational pull of Uranus, causes the interior of Miranda to flex. This tidal flexing may be the cause of the heat leading to the coronae and surface fractures. The gravitational interaction also causes a change in the orbit of Miranda.

A surface temperature of 86 kelvins, determined by Voyager 2, may be enough to melt an ammonia-water mixture. Near-infrared spectrometry not only shows that the surface contains a large amount of water ice; there is also a feature in the spectra that may indicate the presence of ammonia hydrate. This mixture would lower the melting point enough that tidal flexing could account for the fracturing and coronae. The other surface material is carbonaceous and/or silicate matter. The density of Miranda (1.25 grams/centimeter$^3$) is close to that of ice (1.0 grams/centimeter$^3$) so there cannot be a large percentage of other material. The rock fraction can only be 33 percent.

The albedo of Miranda was measured by Voyager at 32 percent. Hubble Space Telescope (HST) measured it at a lower value. One question is, therefore, whether the change in value indicates a real difference, meaning that the northern hemisphere imaged by HST is darker than the southern hemisphere imaged by Voyager, or whether there is another explanation for the difference in values.

There are at least four theories about how Miranda and the other satellites were formed. One is the co-accretion model. Solid particles that were a part of the solar nebula were collected by Uranus's gravitational force to form a disk that later coalesced into satellites. The accretion disk model, by contrast, postulates that the gravitational force of Uranus collected material from the solar nebula. This material formed a disk and later collected into a satellite. A third model is the spin-out model, which suggests that some of the material spun out from the planet as it contracted, forming Miranda and the other satellites. The blow-out model theorizes that an Earth-size body struck Uranus, causing Uranus to tilt and ejecting material that formed a disk and later satellites. Some also subscribe to the theory that Miranda was a small body captured by Uranus's gravitational force as it came close to the planet.

## Knowledge Gained

Miranda can be seen with Earth-based instruments, but it is so close to Uranus that it is difficult to study its characteristics. Near-infrared spectra obtained using the United Kingdom Infrared Telescope (UKIFT) revealed much of what we know about the surface composition of Miranda. Earth-based instruments also serve well to study orbits, even for a faraway satellite like Miranda. Once Miranda was known, it was clear that it had been seen in other photographs. Putting all of the sightings together produced a knowledge of the satellite's orbit.

Goals for the Voyager studies for Miranda were twofold: observational and magnetic. The magnetic goals were to see if Miranda had a magnetic field (it was determined that it does not), how Miranda interacted with Uranus's field, and if there were charged particles in the atmosphere. The observational goals for Voyager were to note the satellite's size, shape, mass, density, shape changes with time; types of surface structures; surface processes; and albedo change with phase angle. Voyager data concerning albedo are similar but slightly lower than that obtained from the Hubble Space Telescope. Voyager data and years of observations from Earth were used together to calculate better values for Miranda's orbit, eccentricity, and angle of inclination. The measurement of the mass of Miranda was made from radio science data. Imaging data were used to determine size, which is more complicated than merely measuring distances in a single image. Several images were needed to cover Miranda. By reference to the same feature in at least two images, researchers then generated a mathematical model to calculate where each feature should be. From that model the size could be calculated.

Comprehensive photometry has been done of Miranda using the HST. All possible phase angles from Earth were studied. A phase angle is the angle between the Sun-to-Miranda line and the Earth-to-Miranda line. The geometric albedo and opposition surge were measured for Miranda. Opposition surge is the brighter reflection that occurs at phase angle of zero.

Together, the data from Voyager and the HST not only vastly increased our knowledge of Miranda but also added to what is known of Uranus's satellites in general. Prior to these space-based observations, the number of known satellites of the gas giant was five; it is now known that there are at least twenty-seven Uranian satellites.

## Context

Data gained by a study of Miranda may be second in importance to the new ideas generated by its unusual surface features. Scientists have to develop new explanations for the generation of the features of Miranda. Some information is known, but the total explanation will require innovative ideas.

Ideas about tectonics and how surfaces of worlds can move and change will help scientists to understand the creation of Earth and the separation of the continents. Studies of tidal flexing, such as that which occurs in the Uranian system, may be important to the study of Earth's plates tectonics. The effect of gravity on the tides is visible, but is that all the effect that gravitational pull has on the Earth? Scientists are also learning more about melting points and sublimation values of mixtures as they try to explain Miranda.

The best way to study Miranda will be to mount another interplanetary probe such as Voyager, but until such a mission can be planned, new techniques are being developed for using land-based instruments and the Hubble Space Telescope to study Miranda and other distant worlds. Humanity will have to improve the instrumentation used presently, and new devices and techniques will have to be designed and tested to expand the study of Miranda.

*C. Alton Hassell*

## Further Reading

Bond, Peter. *Distant Worlds: Milestones in Planetary Exploration.* New York: Copernicus Books, 2007. The author discusses each system; its parts, such as planet, moons, and rings; and how each mission has developed our knowledge of that system. Illustrations, bibliography, appendix, index.

Corfield, Richard. *Lives of the Planets.* New York: Basic Books, 2007. The author takes the reader through the different space missions that have contributed to our knowledge of the planets. The book is divided by planets, and the information gathered by each mission is discussed. Index.

Croswell, Ken. *Ten Worlds: Everything That Orbits the Sun.* Honesdale, Pa.: Boyds Mills Press, 2007. Each system is discussed in turn. Written for a younger audience, this book provides good basic information. Illustrations, bibliography, index.

Elkins-Tanton, Linda T. *Uranus, Neptune, Pluto, and the Outer Solar System.* New York: Chelsea House, 2006. This book explores the Sun's relationship with the three outer planets and their moons. It looks at these planets as recorders of the formation of the solar system. Aimed at a general or high school audience. Illustrations, bibliography, index.

Loewen, Nancy. *The Sideways Planet: Uranus.* Mankato, Minn.: Picture Window Books, 2008. An educational children's book devoted to the planet Uranus. Covers the planet's rings, moons, and tilted axis.

McFadden, Lucy-Ann Adams, Paul Robest Weissman, and T. V. Johnson, eds. *Encyclopedia of the Solar System.* San Diego: Academic Press, 2007. The editors have collected articles written by many experts. It is one of the best surveys of material about the solar system. Illustrations, appendix, index.

Miller, Ron. *Uranus and Neptune.* Brookfield, Conn.: Twenty-First Century Books, 2003. Considers Uranus and its satellites in comparison with other gas giants, especially Neptune, including their atmospheres.

Miner, Ellis. *Uranus: The Planet, Rings, and Satellites.* New York: Ellis Horwood, 1990. The author thoroughly covers the topics of both the Uranian system and the Voyager mission. Miranda is featured as a remarkable satellite. Illustrations, bibliography, index.

# NEPTUNE'S ATMOSPHERE

**Categories:** The Neptunian System; Planets and Planetology

*Neptune's atmosphere exhibits surprisingly rapid changes, large cloud systems, extremely fast winds, and large-scale vertical motions associated with heating from outside and inside. These features were quite unexpected for a planet so far away from the Sun, where astronomers expected to find a cold, featureless atmosphere.*

## Overview

Current models for Earth's atmosphere assume that the primary source of energy is solar radiation. Some basic features of a planetary atmosphere can be predicted from the planet's mass and surface temperature. The mass has to be large enough that the escape velocity of gas at the surface is higher than the random thermal motion speed of most of the molecules. The temperature must be neither too high (as at Mercury) nor so low that everything condenses and freezes. Atmospheric pressure at any level can then be related to the weight per unit area of the gas above that level.

Neptune's mass is $1.02 \times 10^{26}$ kilograms, which is roughly 17.09 times that of Earth. It radiates nearly twice as much energy as it receives from the Sun, so that its internal heat is a major factor in the atmospheric dynamics.

The "surface" of Neptune is arbitrarily defined as being the level where the gas pressure reaches 1 bar or 101.325 newtons per square meter, this being the value of standard sea-level pressure on Earth. The planet's radius is defined at this level. At the equator, this radius is 24,764 kilometers. Here the acceleration due to gravity is 10.11 meters per second squared, close to the Earth surface value of 9.81 meters per second squared.

However, the gaseous atmosphere extends down to levels where the pressure exceeds 1,000 bars. The boundary between the condensed (solid and liquid) and gas-state substances is ill defined but is believed to occur at about the same radius as the Earth's surface radius, which is only about one-quarter of the radius of Neptune. Substances such as hydrogen, helium, and methane exist as solids called "ices," even though their temperature is on the order of 700 kelvins or more.

The temperature decreases outward from about 130 kelvins at 10 bars to 59 kelvins at 1 bar, the arbitrary gaseous "surface" of zero altitude. At altitudes where the pressure drops from about 3 bars to 1 bar, the temperature lapse rate on Uranus, which is similar in many respects to Neptune, is on the order of 1 kelvin per kilometer. The temperature drops to a minimum of around 50 kelvins at an altitude of roughly 50 kilometers, where the pressure is 0.1 bar. This is taken as the tropopause, or top of the troposphere. The temperature then rises almost linearly to about 150 kelvins at about 200 kilometers, where pressure is 0.5 millibar. Temperature remains constant at this level to about 500 kilometers, where pressure is 0.001 millibar, defining the stratosphere. It then rises again, reaching 275 kelvins at roughly 800 kilometers, where pressure is $1 \times 10^{-5}$ millibar. This is the mesopause. At these high levels, the temperature is believed to be due to any of three mechanisms: the "dayglow," where diatomic hydrogen gets dissociated by collision with low-energy electrons;

*Neptune, with its high-altitude clouds, from Voyager 2.* (NASA/JPL)

ammonia crystal clouds form at around 5 bars and 120 kelvins. An ocean of ammonia dissolved in water has been postulated to exist below the 5-bar region. Frozen methane clouds occur at around 2 bars to 0.5 bar (70 kelvins to around 55 kelvins), according to predictive models. The temperature at the cloud tops is approximately 55.1 kelvins. The fraction of methane in the atmosphere at the 1-bar level is about 2 to 3 percent.

Absorption of infrared by methane contributes to the observed light blue color of Neptune. The fraction of methane drops significantly at higher altitudes but is still well above the ratio of methane to hydrogen found in the atmosphere of the Sun. Traces of carbon monoxide (CO) of about 1 part per million and hydrogen cyanide (HCN) of about 1 part per billion have been detected. The HCN is supposed to be formed through photochemical reactions with molecular nitrogen in the upper atmosphere. The abundance of CO in the stratosphere is greater than what can be explained by photochemical formation. One hypothesis is that CO comes from the lower regions. Another is that CO is formed when ice-bearing meteorites ablate in the upper atmosphere, with the methane participating in the reaction.

Neptune orbits the Sun once every 165 years but rotates on its axis once every 16.05 hours. The axis is inclined at 29°, so there are noticeable seasonal changes. Over the past few decades, the south pole has been closer to the Sun and is hence warmer. The cloud cover has become noticeably brighter, and there is evidence of strong updrafts.

Heidi Hammel and other researchers have concluded from analysis of narrow-angle images from Voyager 2 that large-scale cloud features in the equatorial and tropical regions move fast enough against the direction of rotation that their period of rotation about the axis is as high as 18.4 hours, compared to the planet's 16.05-hour day, which is deduced from radio signals based on the planet's internal rotation. This gives wind speeds of 325 meters per second, or 1,170 kilometers per hour. A large "Dark Spot" seen in the southern atmosphere by Voyager 2 appears to have been replaced by several dark spots in the

heating by production of auroras due to interaction of the magnetic field with ions; or joule heating as ions are accelerated by the magnetic field, relative to the movement of the neutral gases in atmospheric winds.

The average composition of Neptune's atmosphere is 77 to 83 percent hydrogen, 16 to 22 percent helium and 1 to 2 percent methane, with about 0.019 percent hydrogen deuteride and traces of ethane. Oxygen is present in water, nitrogen in ammonia, and carbon dominantly in methane, though some carbon monoxide has been detected on Neptune. Other elements, such as sulfur, have also been detected. It is believed that in the molten core of the planet there is silicate rock.

Condensation clouds are present at various levels in the Neptune atmosphere. Clouds of water, ammonia, and hydrogen sulfide are predicted to form with a base at a level where pressure is more than 500 bars and the temperature around 480 kelvins. Around 75 bars and 320 kelvins, "solution" clouds are present. At around 50 bars and 270 kelvins, water-ice clouds are present. Between 2 and 5 bars (100 and 150 kelvins respectively), clouds of solid particles of ammonium hydrosulfide form. Frozen south and later in the far northern hemisphere. These appear to be storms that form and dissipate. In the dark spots, wind speeds up to 2,400 kilometers per hour have been postulated. Several bright spots seen in infrared images appear to indicate warmer rising plumes of air that come from warm layers deep down but reach the upper atmosphere.

## Knowledge Gained

Much of what is known about Neptune comes from the Voyager 2 spacecraft, which transmitted images from an approach distance of 4.48 million kilometers in 1989. The Hubble Space Telescope in 1994-1995, mid-infrared data from the Keck telescope in Hawaii, the National Aeronautics and Space Administration (NASA) Mauna Kea telescope, and the Very Large Array radio telescope in New Mexico have added to our understanding of Neptune.

Predictive models use data from the observed emission and absorption through the atmosphere at various wavelengths of radiation. Sunlight reflected off clouds in the upper troposphere and lower stratosphere shows bright

*Voyager 2 images dark spots on Neptune; three are visible here, including one in the south with a white center.* (NASA/JPL)

bands between 20° and 50° south. Distinct bright clouds are seen around 70° south. A bright south polar "dot" is visible. Observations in narrow bands corresponding to the absorption wavelengths of methane and ethane show that in the south polar regions, these substances appear to be less abundant. Scientists argue that this indicates a subsiding flow in the south polar region that draws the light gases down into warmer regions of the atmosphere and prevents them from freezing, as would happen if they rose into the upper atmosphere. The subsiding flow heats the lower levels adiabatically, while the lack of frozen crystals in the "dry" upper atmosphere above the south polar region makes the atmosphere transparent to much greater depths and allows the warmer regions below to be detected. It appears that air is rising more in the mid-southern and northern latitudes, and sinking at the equator and the south pole. This global circulation pattern is partially attributable to solar heating. Smaller clouds, such as the bands observed near 70° south latitude, are attributed to local weather.

A controversial theory for the intense atmospheric activity seen on this distant planet comes from Glenn Orton and his team, investigating an "electric universe theory." They have reported temperatures near Neptune's south pole that are high enough to allow gaseous methane from lower levels to escape into the upper atmosphere without freezing. They attribute this level of energy input to Neptune to an electrical connection with the rest of the solar system and out to the surrounding interstellar environment. Thus the heating is associated more with perturbations of the magnetic field of Neptune than with direct optical flow of thermal energy from the faraway Sun. According to this theory, Neptune is part of the "electric circuit" and hence can experience levels of energy input that are impossible to predict using just the distance from the Sun. Orton and his colleagues argue that conventional models of atmospheric dynamics based on solar heating fail when dealing with faraway planets.

## Context

Expected to be much colder and less active than Uranus because it is 30 AU from the Sun, Neptune has surprised scientists with its profound seasonal changes in cloud cover and atmospheric circulation over the past twenty years. This is about half of a season. Hammel and colleagues conclude that cloud-top wind speeds are roughly the same order for all planets ranging from Venus to Neptune, even though solar energy inputs to their atmospheres differ by three orders of magnitude. Estimates of Neptune's internal energy and the amount of solar radiation that reaches it do not yet explain this level of activity. Radical theories of electric current flows from the Sun to the outer limits of the solar system have been advanced as alternative explanations. Neptune demonstrates, therefore, how the study of an extreme situation can challenge models based on Earth's atmospheric behavior.

Continued observation of Neptune with both ground-based telescopes and the Hubble Space Telescope must suffice until a Neptune orbiter similar to either the Galileo or Cassini designs might be dispatched to Neptune. In early 2002, such a probe, called the Neptune Orbiter Probe, was investigated but not funded—in part because of its projected cost and in part because it would have required nuclear propulsion to make a trip to Neptune in a reasonable amount of time. Such propulsion technology was not yet available. When the James E. Webb Space Telescope, an infrared observatory, enters service, it will join the instruments currently used to scrutinize the dynamics of Neptune's atmosphere.

*Narayanan M. Komerath*

### Further Reading

Cruikshank, Dale P. *Neptune and Triton*. Phoenix: University of Arizona Press, 1995. Chapters are based on papers presented at an international conference of researchers and span all aspects of the Neptunian system based on Voyager images and data.

Encrenaz, Thérèse, et al. *The Solar System*. New York: Springer, 2004. A thorough exploration of the solar system, from early telescopic observations through the space missions that had investigated all planets (with the exception of Pluto) by the publication date. Takes an astrophysical approach to place the solar system in a wider context as just one member of similar systems throughout the universe.

Hammel, H. B. *The Ice Giant Systems of Uranus and Neptune*. New York: Springer, 2006. An authoritative discussion of the state of our knowledge of the Neptunian atmosphere and other aspects of both Uranus and Neptune, including the rings and moons of these planets.

Irwin, Patrick G. J. *Giant Planets of Our Solar System: An Introduction*. 2d ed. New York: Springer, 2006. Suitable as a textbook for upper-level college courses in planetary science, this volume focuses on Jupiter, Saturn, Uranus, and Neptune and their satellites, rings, and magnetic fields. Filled with figures and photographs. Accessible to serious readers.

Miner, Ellis D., and Randii R. Wessen. *Neptune: The Planet, Rings, and Satellites*. New York: Springer, 2002. Authors were members of the Voyager team during the Neptune encounter. This volume is another member of the Wiley-Praxis series in astronomy and astrophysics and is accessible to the general reader interested in planetary science.

# NEPTUNE'S GREAT DARK SPOTS

**Categories:** The Neptunian System; Planets and Planetology

*An elliptical atmospheric feature in Neptune's southern hemisphere—one large enough to contain Earth—was named the Great Dark Spot in 1989. It was interpreted as an anticyclonic storm. This dark spot eventually disappeared, but others subsequently appeared in the high northern latitudes. Given our understanding of terrestrial storms as driven by solar heating, and that Uranus, which is larger and closer to the Sun, did not show such activity until recently, such phenomena on distant Neptune were very surprising.*

### Overview

Images taken through filters that permit the passage of light at wavelengths of 467 nanometers were taken by the Voyager 2 spacecraft in 1989 at a distance of 2.8 million kilometers. Those images revealed an elliptic feature darker in albedo by about 10 percent. Covering between $30°$ and $45°$ of longitude and $8°$ to $17°$ of latitude, the Great Dark Spot (GDS) is about the size of Earth. It was initially positioned at $27°$ south latitude, drifting toward the equator at about $1.2°$ per month. Given the nomenclature GDS89, the Great Dark Spot was tracked from $27°$ to $17°$ south latitude over eight months and interpreted as a depression in Neptune's atmosphere, indicating a strong vortex such as a hurricane. GDS89 had a companion bright cloud band on one edge. This surprised astronomers after receiving featureless images of Uranus, a planet larger and closer to the Sun, during the Voyager 2 flyby in January, 1986.

In order to understand the concept of the Great Dark Spot, it is helpful to understand the vortex dynamics of storms. When gas heats up in the lower atmosphere and becomes less dense, it forms a "sinklike" flow, moving inward and then rising. The plume starts rotating, since conservation of angular momentum amplifies any difference in tangential speed as the radius decreases. In hurricanes that cover many degrees of latitude on a planet, the direction of rotation is predictable because of the Coriolis effect, whereby wind velocity interacts with the planet's rotation. The core remains clear of clouds and has rising warm air inside, while clouds revolve fastest around its periphery. As the plume reaches levels with lower density, the flow spreads out and the rotation slows outward. The warm, moist air (above Earth) condenses into clouds. Outside the core, cold air sinks. Thus, from observed cloud-top movement and temperature gradients, the strength of the storm and its axial upflow can be determined, along with the density differences and energy input that drive the storm. Prevailing regional winds and the Coriolis effect drive the storm across longitudes and toward the equator.

Near GDS89, wind speeds up to 2,400 kilometers per hour were recorded, but prevailing regional winds were on the order of 900 to 1,500 kilometers per hour. In addition to its drift, GDS89 showed oscillations of about 14° amplitude about the horizontal, with an eight-day period. Its aspect ratio fluctuated between 0.35 and 0.55, the longitudinal size varying between 30° and 45°, and the latitudinal extent between 12° and 17°. Over a 225-day period, GDS89 became increasingly circular. It also showed "tadpole-like tails," dark features off each of the smaller sides. A 1990 paper reported a dynamic model for such oscillations based on a single isolated vortex embedded in a shearing flow, and derived lower limits for the Rossby radius of deformation relating Coriolis forces and buoyancy forces.

In 1994, the Hubble Space Telescope (HST) observed that GDS89 had disappeared, but other dark spots appeared in the northern hemisphere. The Northern Great Dark Spot (NGDS32), a very stable dark spot, stayed near 32° north from around 1994 to 1996, and perhaps until 2000. Another dark spot, NGDS15, stayed near 15° north from March, 1996, to June, 1997. NGDS32 drifted across longitudes very steadily at about 36° per day. The drift rate of NGDS15, being much closer to the pole, proved harder to estimate. Bright methane clouds were

associated with specific latitudes in any given period, but the active latitudes changed from -25° and -30° in 1989 to -30° and −46° in 1994-1996. Astronomers Heidi Hammel and G. Wesley Lockwood noted that GDS appeared in images where the dominant radiation was at 467 nanometers (blue), close to the brightest spots in the red (619 nanometers) and infrared (889 nanometers) existing on the planet at the time.

More detailed three-dimensional vortex simulations in 1998 provided explanations for the appearance of overlapping ellipses and correlated the drift rate of the spots with the prevailing wind speeds in the region. They predicted the breakup of these anticyclones near the equator, with Rossby waves propagating out over a few weeks. A 2001 paper in *Icarus* by P. W. Stratman et al. used simulations of storms, along with measurements of the bright accompanying clouds of the GDS, to estimate the atmospheric level where the top of the GDS should occur. If the top were in the stratosphere, the GDS would drift too quickly toward the equator and disperse. If the top were deep inside the troposphere, the clouds would be much larger than what was seen. Hence the top of a GDS should be near the tropopause. Based on this result, the pressure drop along a streamline threaded through a companion cloud was on the order of 3 millibars and the temperature change was on the order of one kelvin, indicating a lifting on the order of half a kilometer and relative wind speeds between the dark spot and the surrounding winds of 45 meters per second eastward.

However, a 2002 paper in *Icarus* by L. A. Sromovsky et al. showed that the steady latitudes maintained by NGDS32 and NGDS15 are not consistent with the model of anticyclonic storms, which should have moved strongly across latitudes. Computational fluid dynamic simulations by R. P. LeBeau and colleagues in 2006 and 2007 captured shape and oscillation phenomena similar to those of GDS89, but the size and oscillation amplitudes were off by as much as a factor of two. They calculated the shear profile in the background winds of Neptune that would be needed to explain the slow drift toward the equator of GDS89, but the much slower rates of the northern dark spots remain a challenge to explain.

In comparison, for more than three hundred years the Great Red Spot (GRS) of Jupiter has circled that planet along a southern latitude with minimal longitudinal drift. Attempts to explain the GRS as the flow around a high solid surface peak have been abandoned, and it is now considered to be a shallow cloud system trapped between shearing layers of horizontal winds. The redness relative to the surrounding white ammonia clouds indicates some temperature difference. Efforts to model the GRS as a vortical upwelling of fluid from below have met with limited success. It is not certain that the dark spots of Neptune are shallow structures, or that they extend to the cloud tops, rather than being features lying below clear atmospheric regions.

In 2006, the Hubble Space Telescope detected a dark spot more than 1,000 kilometers in extent, at 27° south latitude on Uranus. This observation was made as that planet began experiencing increased atmospheric activity with the coming of summer in its eighty-four-year orbit around the Sun. Scientists believe that Uranus is not as bland as Voyager 2's images suggested. Instead, as the amount of solar radiation that the planet intercepts increases, the planet may develop features seen in the atmospheres of the other gas giant planets.

## Knowledge Gained

What is known about the Great Dark Spots comes from images taken by the Voyager 2 spacecraft in 1989, by the Hubble Space Telescope from 1994 to 2000, and by ground-based optical and radio telescopes, including the Keck telescope, the Mauna Kea observatory, and the Very Large Array radio telescope in New Mexico. These are all passive observations, and the investigations for which

*Voyager 2's close-up image of Dark Spot 2, with its white central structures, suggesting clouds.* (NASA/JPL)

they provide evidence depend on analyzing images taken with various filters that show specific wavelengths. Given the great distance, even the latest Hubble observations do not begin to approach the resolution achieved with Voyager 2's 1970's-vintage cameras. The narrow-angle camera on Voyager 2 was the instrument used to capture cloud-top images, which were then used to calculate wind speeds at that level.

Radio telescopes capture signals that should indicate rotation of the magnetic field and the internal structure of the planet, and hence give rotation rates based on those factors. With a planet composed mostly of fluid, there can be large differences between these rotation rates, which remain unexplained but are attributed to extremely high winds, which imply large frictional losses that require high energy input. The structure of the dark spots is derived mostly from simulations of fluid mechanics based on what is known of terrestrial storm systems and the limited data from these planets.

Researchers continue to model the dark spots using the fluid mechanics of hurricanes. Their speed of travel around the planet appears to match what is known or assumed of local wind speeds. However, their slowness in crossing latitudes requires fortuitous combinations of wind profiles to explain. Their shorter persistence compared to the Great Red Spot of Jupiter appears consistent with the much higher "wind speed" and the shear between different zones prevailing on Neptune. The spectral contrast suggests the presence of different gases and different temperature in the center of a spot, suggesting strong vertical motions of gas from the warmer depths of Neptune. The absence of features inside the dark spots, unlike the clouds seen above the GRS, frustrates efforts to derive their interior structure and vorticity directly.

## Context

The driving engine for Neptune's dark spots remains puzzling, since the small changes in solar intensity associated with seasonal changes do not provide sufficient differences to explain these mysterious phenomena. If such strong weather activity can occur with such low intensity of sunlight, clearly much remains to be learned about why the Earth's weather behaves as it does. Observations of summertime on Uranus in the next decade may explain some of the mysteries, but Neptune's great dark spots are unlikely to yield their secrets until spacecraft descend through the Neptunian atmosphere to probe the planet's winds. New discoveries from such missions could greatly improve our ability to predict the course and evolution of killer storms on Earth.

In 2002 a study was convened to investigate the possibility of a Neptune Orbiter Probe, but the concept was not funded and was hampered by the need for nuclear propulsion technology, still to be developed. Until the development of a nuclear propulsion system capable of making the journey to Neptune in a timely fashion with a large payload, the likelihood of a follow-up spacecraft-based investigation of Neptune is relatively small. In the meantime, continuing study of Neptune and its atmospheric features will be conducted using the Hubble Space Telescope and ground-based observatories like the Keck telescope.

*Narayanan M. Komerath*

## Further Reading

Cruikshank, Dale P. *Neptune and Triton*. Phoenix: University of Arizona Press, 1995. Chapters are based on papers presented at an international conference of researchers and span all aspects of the Neptunian system based on Voyager images and data.

Encrenaz, Thérèse, et al. *The Solar System*. New York: Springer, 2004. A thorough exploration of the solar system, from early telescopic observations through the space missions that had investigated all planets (with the exception of Pluto) by the publication date. Takes an astrophysical approach to give our solar system a wider context as just one member of similar systems throughout the universe.

Freedman, Roger A., and William J. Kaufmann III. *Universe*. 8th ed. New York: W. H. Freeman, 2008. A college text on astronomy, somewhat more advanced than many introductory texts, with a wealth of detail and excellent diagrams. Chapters 6 through 16 describe the solar system. Comes with a CD-ROM.

Hammel, H. B., and G. W. Lockwood. "Atmospheric Structure of Neptune in 1994, 1995, and 1996: HST Imaging at Multiple Wavelengths." *Icarus* 129, no. 2 (October, 1997): 466-481. This paper uses the Hubble telescope to relate observations of dark spots on Neptune to the brightness of methane clouds. It shows that dark spots occur near the region where the brightest features appear in the red (619-nanometer) and near-infrared (889-nanometer) bands.

Hammel, H. B., G. W. Lockwood, J. R. Mills, and C. D. Barnet. "Hubble Space Telescope Imaging of Neptune's Cloud Structure in 1994." *Science* 268, no. 5218 (1995): 1740-1742. Discusses image results, wind speeds, the disappearance of the original dark spot, and the appearance of new ones. Also discusses methods used to filter the images to bring out different features based on the absorption of red and infrared by methane, and the bright scattering from high-altitude clouds above the gaseous methane layers.

Irwin, Patrick G. J. *Giant Planets of Our Solar System: An Introduction*. 2d ed. New York: Springer, 2006. Suitable as a textbook for upper-level college courses in planetary science. Focuses on Jupiter, Saturn, Uranus, and Neptune and their satellites, rings, and magnetic fields. Filled with figures and photographs. Accessible to serious readers.

Miner, Ellis D., and Randii R. Wessen. *Neptune: The Planet, Rings, and Satellites*. New York: Springer, 2002. The authors were members of the Voyager team during the Neptune encounter. Accessible to the general reader interested in planetary science.

Oxlade, Chris. *Jupiter, Neptune, and Other Outer Planets*. New York: Rosen Central, 2007. Intended for middle school students, this book compares and contrasts the Jovian planets using the latest data available on the gas giants and their features.

# NEPTUNE'S INTERIOR

**Categories:** The Neptunian System; Planets and Planetology

*Neptune is a gas giant planet, so its interior structure is completely different from that of the terrestrial planets and most similar to that of Uranus, another gas giant. The most likely structure consists of a relatively small rocky core, an icy layer, and a molecular hydrogen layer. The interior is then covered with an atmospheric layer of hydrogen, helium, and various other gases. Neptune's interior lacks the metallic hydrogen layer found in Jupiter and Saturn.*

## Overview

It is not possible to conduct direct studies of planetary interiors, including Earth's, so astronomers must resort to indirect clues to infer the interior structures and compositions of planets. These clues can include characteristics of the bulk density, the planetary magnetic field, seismic activity and waves for terrestrial planets, and heat flow from the interior.

The two major classes of planets in the solar system are the Jovian planets and the terrestrial planets. Jupiter is the prototype for the Jovian planets, which also include Saturn, Uranus, and Neptune. Earth is the prototype for terrestrial planets, which also include Mercury, Venus, and Mars. Earth's moon is also very similar in structure to the terrestrial planets. In oversimplified terms, Jovian planets are big balls of gas, and terrestrial planets are small balls of rock and metal.

A planet's bulk density is one clue to its interior composition and structure. Density is defined as the mass divided by the volume. Both an object's mass and its volume depend on the object's composition and how much of that material there is. For example, a boulder has a greater volume and mass than a pebble made of the same type of rock. However, the amount of material cancels out when dividing the mass by the volume. Therefore the density of a boulder and the density of a pebble made of the same type of rock will be the same, even if the boulder is as large as a planet. Density is a property of the type of material but not how much material there is, so the density of a planet can tell us something about its composition. Water has a density of 1.000 kilograms per meter cubed (km/m³). Planets with approximately this density are primarily icy materials or gas. Planets with a density of approximately 2,000 to 3,000 km/m³ are typically made of rocky materials. Planets with a density of 7,000 to 9,000 km/m³ would be primarily metallic in composition. Planets with densities between these ranges are mixtures of materials. Neptune has a density of 1,600 km/m³. It must therefore be mostly ice or gas. This density, the highest of the Jovian planets, is slightly greater than the density of Uranus and significantly greater than the densities of Jupiter and Saturn. Therefore, the rocky cores of Neptune and Uranus constitute a larger portion of their total mass than do the cores of Jupiter or Saturn. Because Neptune's density is slightly greater than that of Uranus, the two planets must have differences in their interior compositions, despite their very close similarities.

Seismic activity can give us clues to the interiors of only the terrestrial planets (which are solid and therefore have seismic activity). Neptune is a gas giant planet with no potential for seismic activity, so this potential clue does not apply to Neptune.

Because the Jovian planets are gas, astronomers can use the ideal gas law for theoretical calculations of interior properties. This equation relates the temperature, density, and pressure of the gas.

Planetwide magnetic fields give us clues to the planets' interiors. All magnetic fields are ultimately produced by some type of electric current, so in order for a planet to have a magnetic field, its interior must contain a liquid or gas layer that can conduct electricity. The planet's rotation helps set up the electric currents in this layer. A composition of iron or other ferromagnetic materials can enhance the magnetic field produced by the interior electric currents but is not essential. Neptune does have a strong magnetic field that is comparable in strength to those of Saturn and Uranus. Like that of Uranus, Neptune's magnetic field axis is tilted significantly from its rotational axis. The tilt is 46° in the case of Neptune. The center of Neptune's magnetic dipole is displaced significantly from the center of the planet. This displacement is even greater than for Uranus. No other planets have such displacements or tilts of their magnetic dipoles. If the icy layer is not completely solid, but at least partially molten, then the water can conduct electric currents and produce a magnetic field. Impurities in the water increase its ability to conduct electricity. Hence we infer that Neptune's icy layer is not completely solid and the molten regions are not symmetric about the center of the planet.

With the exception of Uranus, the Jovian planets emit more energy than they receive from the Sun. Neptune emits about three times as much energy as it receives from the Sun. This extra energy must come from somewhere, and the only place available is the planet's interior. Therefore the interiors of the Jovian planets, including Neptune, must be hot. When a gas giant planet forms, its own gravity compresses the gas. A gas heats up when it is compressed, so the interiors of gas giant planets should be hot. Neptune therefore has a hot interior, as expected from the behavior of gases and confirmed by the heat flowing outward from the interior to the surface. The existence of large storms and weather on Neptune provides additional evidence that heat energy, which is needed to power storms, flows upward from the interior.

Putting all these clues together, astronomers can infer the interior structure of Neptune. The innermost layer is a rocky core. Neptune's rocky core is approximately the size of Earth and has a temperature of about 7,000 kelvins. The mass of Neptune's rocky core is most likely a little less than ten times the mass of Earth, but estimates range from four to fifteen times Earth's mass. The rocky core is more massive than Earth, even though it is about the same size, because the high temperatures and pressures in the core of the planet increase the density of the rock.

Above the rocky core, Neptune has an icy mantle consisting primarily of water, methane, and ammonia ice that is not completely frozen. The ice is slushy rather than completely frozen. Hence convection currents can flow in this mantle. The convection currents carry heat from the interior to the surface. They also help form Neptune's magnetic field. As the slushy material circulates, it is moving in the Sun's extended magnetic field. This motion induces electric currents in the slush. The slushy water ice has ammonia dissolved in it, so it conducts electricity very well. These induced electric currents induce Neptune's magnetic field. Electric currents induce magnetic fields and moving or changing magnetic fields induce electric currents. If the regions where the magnetic field is generated are far from the center of the planet and distributed asymmetrically around the planet, then the magnetic field would be tilted and displaced as the Voyager 2 data show.

The outermost layer of Neptune's interior is molecular hydrogen. The temperature, pressure, and density of the molecular hydrogen layer increases with depth. The icy layer can form despite high temperatures because the freezing temperature of water increases as the pressure increases.

Neptune's interior structure is similar to that of Uranus but dissimilar from the interiors of the other gas giants, Jupiter and Saturn, which have a metallic hydrogen layer in addition to these layers. Neptune and Uranus do not have metallic hydrogen layers because the pressure in the hydrogen layers is not high enough for hydrogen to become metallic hydrogen rather than molecular hydrogen.

## Knowledge Gained

Most of what is known about Neptune and its interior comes from the Voyager 2 mission. After flying by Jupiter and Saturn, Voyager 1 flew out of the plane of the solar system. Voyager 2 flew on to Uranus and Neptune. The Neptune flyby was in 1989. To date, this is the only spacecraft to explore either Uranus or Neptune. All other observations are from ground-based or orbiting telescopes. Examples of the knowledge of Neptune gained from the Voyager mission include measuring the planet's magnetic field, taking more accurate measurements of Neptune's mass and therefore its density, and discovering its surface storms.

The Great Dark Spot that the Voyager discovered on Neptune is a large storm, similar in nature to the great red spot on Jupiter. Since the Voyager mission, the Hubble Space Telescope has been able to take pictures with enough resolution to allow astronomers to follow weather patterns on Neptune. Storm systems on Neptune change. Weather and large storm systems require energy to drive them. On Earth the energy comes from the Sun, but Neptune is too far from the Sun for solar energy to drive the storms and weather astronomers observe. Therefore the needed energy must come from the interior of the planet. Uranus, though very similar to Neptune, does not have such storms. It is deduced, therefore, that Neptune has a much hotter interior than Uranus.

The interior differences among the four Jovian planets provide with clues to their formation. Jupiter, Saturn, Uranus, and Neptune all have rocky cores of approximately the same size and mass. Jupiter and Saturn, however, have much more hydrogen and helium surrounding their rocky cores. This observation suggests that Uranus and Neptune formed much later than Jupiter and Saturn, at a time when the protoplanetary disk had dissipated more of its hydrogen and helium gas. At the greater distance from the Sun, the small planetesimals that had to merge to form the cores of Uranus and Neptune were farther apart and therefore took longer to merge.

Neptune, like the other planets, is differentiated into layers. The denser materials are closer to the center. Differentiation in solid terrestrial planets tells us that they were at one time liquid, because denser materials would sink only in a fluid, not in a solid. Smaller satellites that were never liquid are not differentiated.

## Context

Comparing the interiors of other planets casts light on Earth's interior, just as knowledge of Earth's interior provides insights into other planets. These comparisons of how planetary interiors develop provide clues to the origins of the planets and to the processes that shape planetary interiors and compositions.

To date, only one spacecraft, Voyager 2, has studied the outer planets Uranus and Neptune. No probes have yet flown by Pluto. This paucity of robotic probes to the outer solar system limits our understanding of these planets. When humans send additional spacecraft to the outermost reaches of the solar system, our understanding will increase. When we send another probe to Neptune, our knowledge of the planet and its interior will increase dramatically. It is very likely to completely change our ideas about Neptune's interior.

*Paul A. Heckert*

## Further Reading

Chaisson, Eric, and Steve McMillan. *Astronomy Today.* 6th ed. New York: Addison-Wesley, 2008. One chapter of this readable introductory astronomy textbook covers Neptune, Uranus, and Pluto.

Freedman, Roger A., and William J. Kaufmann III. *Universe.* 8th ed. New York: W. H. Freeman, 2008. Chapter 14 of this introductory astronomy textbook is a complete overview of the outermost planets in the solar system.

Hartmann, William K. *Moons and Planets.* 5th ed. Belmont, Calif.: Thomson Brooks/Cole, 2005. This textbook on the satellites and planets of the solar system summarizes our understanding of planetary interiors in chapter 8.

Hester, Jeff, et al. *Twenty-First Century Astronomy.* New York: W. W. Norton, 2007. Chapter 8 of this well-illustrated astronomy textbook is about the Jovian planets and includes a good discussion of their planetary interiors.

Morrison, David, Sidney Wolf, and Andrew Fraknoi. *Abell's Exploration of the Universe.* 7th ed. Philadelphia: Saunders College Publishing, 1995. The outer planets are covered in chapter 16 of this classic astronomy textbook.

Zeilik, Michael. *Astronomy: The Evolving Universe.* 9th ed. Cambridge, England: Cambridge University Press, 2002. An extremely well written introductory astronomy textbook. Chapter 10 is an overview of our knowledge of the Jovian planets.

Zeilik, Michael, and Stephen A. Gregory. *Introductory Astronomy and Astrophysics.* 4th ed. Fort Worth, Tex.: Saunders College Publishing, 1998. Aimed at undergraduate physics or astronomy majors, this textbook goes into more mathematical depth than most introductory astronomy textbooks. Chapter 6 covers the basic principles of the Jovian planets.

# NEPTUNE'S MAGNETIC FIELD

**Categories:** The Neptunian System; Planets and Planetology

*Neptune is the eighth planet outward from the Sun. Although classified a Jovian or "gas giant" planet, Neptune shares more similarities with Uranus than it does with Jupiter or Saturn. Neptune's magnetic field is inferred to be generated in a manner akin to that which produces Uranus's magnetic field.*

## Overview

The first planet to be discovered during the telescope age was Uranus, the seventh planet from the Sun. The planets from Mercury (closest to the Sun) out through Saturn (sixth from the Sun) were known to the ancients. No records exist that pinpoint a discovery event in each of these planets' cases. After Sir William Herschel announced the discovery of Uranus in 1781, observers began making precise determinations of Uranus's orbit about the Sun.

Unexpected motions in the orbit of Uranus led many to suspect that a planet existed farther out in the solar system, one that exerted gravitational influences on Uranus that perturbed its orbit. A search followed for a new planet to join the family of seven planets in the contemporary solar system model. This search was greatly guided by mathematical analysis using the young field of celestial mechanics, as based on an understanding of Newtonian gravitation. The basis of that analysis involved calculating where a body would have to be located beyond Uranus in order to account for measured variations in the orbit of Uranus.

Who actually discovered Neptune has been a matter of debate. Calculations in 1843 by John Couch Adams were dispatched to Astronomer Royal Sir George Airy, but the latter failed to be convinced. He even asked Adams to send proof of the validity of his work. Adams appeared to be either disappointed or insulted by Airy's dismissal, and he never responded. Two years later, Urbain Le Verrier published an independent calculation that did generate some interest in Airy. Efforts at Cambridge Observatory failed to find the eighth planet, but when Le Verrier asked the Berlin Observatory to aim a telescope toward a particular region of the sky, Johann Gottfried Galle reacted immediately on the request. Neptune was found on September 23, 1846. Its position was just one degree off from Le Verrier's calculation and twelve from Adams's. In the aftermath, the British and French argued over who actually discovered the new planet. Adding to the controversy was the fact that Cambridge records indicated Neptune was charted months earlier than Galle's observation. However, those records did not recognize Neptune as a planet. Even in the last years of the twentieth century, there were those who argued against giving Le Verrier and Adams joint credit as Neptune's discoverers.

In the thirteen decades since its discovery, information about Neptune has been garnered only by using ever-increasingly large Earth-based telescopes. Then, in 1977, a special opportunity arose for sending a single spacecraft to explore the outer solar system making use of gravity assists in turn from Jupiter to Saturn to Uranus to Neptune.

The National Aeronautics and Space Administration (NASA) was able to launch Voyager 2 on such a "Grand Tour." Voyager 2 encountered Jupiter in 1979, Saturn in 1981, Uranus in 1986, and then Neptune in 1989.

Just as it had at Uranus, Voyager 2 would answer the very basic question of whether or not Neptune had a magnetic field. The answer, be it in the affirmative or the negative, would provide considerable insight into the interior structure of Neptune. Voyager 2 indeed found Neptune to be a dynamic world that generates more heat energy than the radiation it receives from the Sun. It generates a complex and dynamic magnetic field that has left scientists with much to consider concerning the implications of that magnetic field for the structure and evolution of the planet.

## Knowledge Gained

There are a number of means whereby a magnetic field can be detected. Most of them make use of electromagnetic induction in that a voltage is induced in a coil by intercepting a time-dependent magnetic flux within that coil. That is the basis of sophisticated magnetometers such as that flown on the Voyager 2 spacecraft. Scientists had to be careful in interpreting readings from the spacecraft's magnetometer, in that it picked up magnetic effects created by the solar wind even at the distant position of Neptune. Solar wind particles become trapped within a planet's magnetosphere, and that generates its magnetic field.

The Voyager 2 Neptune encounter began on June 5, 1989, with the spacecraft still 117 million kilometers distant from the planet. Voyager first detected radio waves, indicating it had crossed into Neptune's magnetosphere, on August 24. The spacecraft came within 4,400 kilometers of Neptune's upper atmosphere on August 25. The spacecraft remained inside Neptune's magnetosphere for a total of thirty-eight hours and noted that Neptune's magnetic "bubble" was affected by both the planet's ring system and its satellites. The planet's magnetosphere exists for 35 Neptune radii on the sunward side and out to 72 radii behind the planet. That varies with the strength of the solar wind. Several days later the Neptune near-encounter phase ended after Voyager passed close to Neptune's large satellite Triton. Triton especially alters the outermost portion of Neptune's magnetosphere. Triton revolves about Neptune in a retrograde fashion. Charged particles within the magnetosphere near Triton also include nitrogen ions that originate from cryogeyser eruptions on this satellite.

Voyager 2 also found radiation belts trapped in the planet's expansive magnetosphere, although charged particle density in Neptune's magnetosphere was found to be less than that at Uranus. The composition of the trapped particles primarily includes protons, electrons, and ionized molecular hydrogen and helium.

Analysis of Voyager 2 data from both the spacecraft's magnetometer and radio astronomy experiment indicated that Neptune's magnetic field varies considerably as the planet rotates in the presence of the solar wind. Neptune's magnetic field is dominated by its dipole character, but it also has a strong quadrupole component. A quadrupole can be understood as if it were two bar magnets oriented at right angles to each other. The field strength near the equatorial region is 1.42 microteslas. Neptune's dipole strength is $2.2 \times 10^{17}$ tesla-meters cubed. Uranus's dipole moment is nearly double that value. The planet's quadrupole moment is due in large part to an offset of the field center from the planet's center. Octupole and higher moments could not be determined accurately.

The center of Neptune's field is displaced from the planet's center by approximately 0.55 planetary radius, which corresponds to 55 percent of the planet's radius. The planet's magnetic field is also tilted significantly, 47° relative to the planet's rotational axis, as was the case at Uranus.

On Earth, trapped particles spiral around magnetic field lines and dip down into the atmosphere over polar regions, creating auroras by exciting atmospheric gases. Voyager 2 detected auroral activity in Neptune's atmosphere, but in large part because of the complexity of Neptune's magnetic field structure, the observed auroral activity was not confined to areas near the magnetic poles of the planet.

Observations of Neptune's magnetic field provided insight into Neptune's interior. Like Uranus, Neptune is believed to have a liquid layer of high electrical conductivity in motion powered by internal heat flow from the core beneath. This results in a dynamo effect from this fluid mantle region, which is composed of water, ammonia, methane, and lesser amounts of volatile substances. Planetary rotation results in complex variations of Neptune's magnetic field.

Neptune is a radio source as a result of its rotation and the fact that it possesses a reasonably strong magnetic field. Rotation of the magnetic field is determined by the observed periodicity in the planet's radio emissions. That provided a better means of determining Neptune's rotation rate than attempting to monitor the time taken for atmospheric features, such as streamers or large storms, to make a compete rotation. A gas giant's atmosphere rotates differentially. The planetary core is involved in producing the magnetic field, and radio emissions led scientists to determine Neptune's core rotates once every 16.11667 hours.

## Context

After Uranus, Neptune was the second planet in the solar system to be discovered by telescopic observations. Just as the search for Neptune originated based on irregularities noticed in the orbit of Uranus, perturbations of Neptune's orbit led to a search for yet another planet in the outer solar system. In 1930, Pluto was discovered by Clyde Tombaugh, but ironically Pluto did not explain Neptune's orbital irregularities. Also, as of 2006, the International Astronomical Union no longer classifies Pluto as a planet, but rather Pluto is now considered to be either a dwarf planet or the first representative of a class of objects called plutoids.

For more than one hundred years since the planet's discovery, the only means to investigate Neptune was by using ever-larger Earth-based optical telescopes. Then, with the advent of the space age and the confluence of a special configuration of planets in the outer solar system that arises only once every 176 years, it became possible to send a spacecraft from Earth: Voyager 2. It traveled in turn to Jupiter, Saturn, Uranus, and Neptune. The Voyager 2 flyby of Neptune generated the best available data and images about this mysterious blue world to that time.

After that Voyager flyby, the Hubble Space Telescope was used in the 1990's and 2000's to make continual observations of Neptune. However, at this time only a new spacecraft mission can advance our knowledge of Neptune's magnetic field. A proposal advanced in 2002 for the Neptune Orbiter Probe was investigated in 2003 as part of NASA's response to the Bush administration's Vision for Space Exploration but was not funded at that time. Planetary scientists continued to indicate strong support for a mission, in the class of the Cassini orbiter, to investigate both Uranus and Neptune. The need to develop nuclear propulsion to get to the outer solar system, however, will need to be met first.

*David G. Fisher*

## Further Reading

Bredeson, Carmen. *NASA Planetary Spacecraft: Galileo, Magellan, Pathfinder, and Voyager*. New York: Enslow, 2000. This book is part of Enslow's Countdown to Space Series. Provides an overview of NASA planetary exploration during the last two decades of the twentieth century. Designed for younger readers but suitable for all nonspecialist audiences.

Cruikshank, Dale P. *Neptune and Triton*. Phoenix: University of Arizona Press, 1995. Chapters are based on papers presented at an international conference of researchers and address all aspects of the Neptune system based on Voyager images and data.

Encrenaz, Thérèse, et al. *The Solar System*. New York: Springer, 2004. A thorough exploration of the solar system, from early telescopic observations through the space missions that had investigated all planets (with the exception of Pluto) by the publication date. Takes an astrophysical approach to place our solar system in a broader context as just one member of similar systems throughout the universe.

Freedman, Roger A., and William J. Kaufmann III. *Universe*. 8th ed. New York: W. H. Freeman, 2008. A college text on astronomy, somewhat more advanced than many introductory texts, containing a wealth of detail and excellent diagrams. Chapters 6 through 16 describe the solar system. Comes with a CD-ROM.

Hunt, Garry E., and Patrick Moore. *Atlas of Neptune*. Cambridge, England: Cambridge University Press, 1994. A combination of telescope and Voyager images of the planet is supplemented by text on historical and scientific background. Summarizes the contemporary understanding of Neptune. Useful to general readers and scientists alike.

Irwin, Patrick G. J. *Giant Planets of Our Solar System: An Introduction*. 2d ed. New York: Springer, 2006. Suitable as a textbook for upper-level college courses in planetary science. Focuses on Jupiter, Saturn, Uranus, and Neptune and their satellites, rings, and magnetic fields. Filled with figures and photographs. Accessible to the serious general audience.

Kerrod, Robin. *Uranus, Neptune, and Pluto*. New York: Lerner Publications, 2000. Presents investigations of the outer solar system from Pioneer 11 through Voyager 2. Intended for a young audience. Not all photographs are clearly identified, but the text is suitable for the general audience.

McBride, Neil, and Iain Gilmour, eds. *An Introduction to the Solar System*. Cambridge, England: Cambridge University Press, 2004. A complete description of solar-system astronomy suitable for an introductory college course as well as nonscientists. Filled with supplemental learning aids and solved student exercises. A Web site is available for educator support.

Miner, Ellis D., and Randii R. Wessen. *Neptune: The Planet, Rings, and Satellites.* New York: Springer, 2002. The authors, who were members of the Voyager team during the Neptune encounter, have written an book accessible to the general reader interested in planetary science.

Oxlade, Chris. *Jupiter, Neptune, and Other Outer Planets.* New York: Rosen Central, 2007. Intended for middle school students, this work compares and contrasts the Jovian planets.

# NEPTUNE'S RING SYSTEM

**Categories:** The Neptunian System; Planets and Planetology

*While all of the gas giant planets have ring systems, Neptune's are unique. They consist of a series of concentric rings and partial arcs. Why Neptune's rings developed differently is still a matter of debate among scientists.*

## Overview

After Uranus's discovery, scientists noticed that it did not behave as expected. They concluded that gravitational effects of a yet unknown planet caused discrepancies noted in Uranus's orbit. In the early 1840's, British astronomer John Adams and French mathematician Urbain Le Verrier independently calculated the unknown planet's orbit and mass. In 1846, German astronomer Johann Galle discovered Neptune within one or two degrees of its predicted location.

It was not until 1984, however, that scientists at two observatories found the first evidence of a ring system around Neptune. On July 22 of that year, Neptune occulted star SAO 186001. Astronomers use the occultation of stars to look for rings and satellites or to measure a planet's exact size. The observatories (which were located in Chile about 100 kilometers apart) recorded a brief occultation lasting just a second. The starlight was reduced by a mere 35 percent when Neptune passed across the line of sight from Earth to the star. Usually this would be considered evidence of a new satellite. In this case, however, to fit with the data, a satellite would only have been only 10 to 20 kilometers in diameter. The astronomers concluded instead that they had discovered a ring around Neptune. However, neither observatory noticed any reduction in starlight on the other side of Neptune.

Neptune's arc-type ring structures remained a mystery until the Voyager 2 spacecraft arrived at Neptune in 1989. The first images from Voyager were taken at a distance of about 20.8 million kilometers during Voyager's approach to Neptune. The photographs showed two arcs around the planet. The first arc was about 48,000 kilometers long and tilted at a 45° angle with respect to Neptune. The inner arc was about 9,600 kilometers wide, at a distance of approximately 51,680,300 kilometers from Neptune's center. Both arcs were found near recently discovered satellites of the planet. Voyager 2 located the partial rings early enough for scientists at the spacecraft's headquarters in California to redirect its camera toward them. Initial photographs showed a differing brightness in the rings, which scientists thought might have resulted from size, density,

*After reaching Neptune in 1989, Voyager 2 took this image of rings around Neptune.* (NASA/JPL)

or rock variations among the ring particles. Saturn's F ring has a similar varying brightness.

On August 23, 1989, Voyager 2 transmitted images back to Earth that solved the partial ring arc mystery. The arcs are actually full rings that encircle Neptune but are not completely visible from Earth. The spacecraft photographed the rings in both forward and backward scattered light. Microscopic particles cannot be seen from Earth, since they are too small to reflect enough sunlight to be seen at such a distance. However, when they were backlit—imaged with the Sun behind them—Voyager 2's cameras were able to detect them as small dust particulates.

In total, Neptune has five rings and partial arcs. The innermost ring, named Galle, is inside the orbit of Neptune's satellite Naiad. This ring is composed mostly of dust, resembling the partial arcs. The Le Verrier ring is next closest to Neptune. It is also the second most prominent of the rings. It is narrow and dusty, located 700 kilometers outside the orbit of Despina. The widest ring, Lassell, extends for 4,000 kilometers. It is one of the less dusty and more complete rings. Its outer edge, which is significantly brighter than the rest, has been named the Arago ring.

The outer and most prominent ring is named after one of Neptune's discoverers, Adams. Orbiting Neptune about 1,000 kilometers outside the path of Galatea, it is narrow and dim in comparison to the bright rings of Uranus and Saturn. The Adams ring also contains five ring arcs: Courage, Liberté, Égalité (1 and 2), and Fraternité.

After Voyager 2 reached Neptune, astronomers knew that the planet had a complete ring system. They still did not know, however, what had caused the arcs of the Adams ring to remain incomplete. In 1991, planetary scientist Carolyn Porco postulated that the arcs were a result of Neptune's signature satellite Galatea. The small satellite orbits Neptune 1,000 kilometers inside the Adams ring. Porco argues that Galatea acts as a shepherding moon, whose gravitational effects keep the matter in the arcs from forming a complete ring. Analysis of data collected by Voyager 2 shows that the arcs "wiggle," or shift positions, by up to 30 kilometers. Porco believes arc distortion occurs at the right speed to be attributed to the small satellite. Critics of this theory point out that ring-arc material would need to have orbits that intersect each other in order to maintain its overall shape. Porco agrees that this would cause

### Rings of Neptune

| | Radius (km) | Radius/ Eq. Radius | Optical Depth | Albedo (×10⁻³) | Width (km) |
|---|---|---|---|---|---|
| Neptune equator | 24,766 | 1.000 | — | — | — |
| Galle (1989N3R) | ~41,900 | 1.692 | ~0.00008 | ~15 | ~2,000 |
| LeVerrier (1989N2R) | ~53,200 | 2.148 | ~0.002 | ~15 | ~110 |
| Lassell (1989N4R*) | ~53,200 | 2.148 | ~0.00015 | ~15 | ~4,000 |
| Arago (1989N4R*) | ~57,200 | 2.310 | — | — | <~100 |
| Unnamed (indistinct) | 61,950 | 2.501 | — | — | — |
| Adams (1989N1R) | 62,933 | 2.541 | ~0.0045 | ~15 | ~50 |
| Arcs in Adams Ring: | | | | | |
| Courage | 62,933 | 2.541 | 0.12 | — | ~15 |
| Liberté | 62,933 | 2.541 | 0.12 | — | ~15 |
| Egalité 1 | 62,933 | 2.541 | 0.12 | ~40 | ~15 |
| Egalité 2 | 62,933 | 2.541 | 0.12 | ~40 | ~15 |
| Fraternité | 62,933 | 2.541 | 0.12 | — | ~15 |

*Notes:* *Lassell and Arago were originally identified as one ring, designated 1989N4R.
*Source:* Data are from the National Aeronautics and Space Administration/Goddard Space Flight Center, National Space Science Data Center.

collisions leading to the inevitable destruction of the arcs themselves. Her model also only partly explains the possible origin of the arcs. It can only determine places where arcs are more likely to develop, not specifically where they already have developed. Because of the lumpiness of the partial rings, Porco and other scientists believe they could once have been a small moon that was destroyed.

In 1999, a group of scientists argued that Galatea cannot be the sole influence causing Neptune's arcs. They studied data collected in 1998 using the Hubble Space Telescope. The largest change in position was of the Liberté arc, which was displaced 1.9° compared to the Égalité arc. The scientists argued that this finding was not a result of varying particle size, given Voyager data that support size conformity within the arc. Their findings, however, do not rule out the shepherding moon theory that relies on the effects of two satellites. Porco and her colleague Fathi Namouni published a paper in 2002 again arguing in favor of their Galatea shepherding-moon model. They believe that Galatea's elliptical orbit keeps the arc particles from having many collisions. By refining the mathematical model of Neptune's ring system, Porco and Namouni proved that Galatea might still be the answer to the arc mystery. More precise data concerning Galatea's orbit and the partial rings are needed before scientists will be able to determine what really causes the arcs around Neptune.

In 2002 and 2003, a group of scientists led by Imke de Pater photographed Neptune's outer rings. Studying the images, they discovered that all of Neptune's arcs appear to be fading away. The Liberté arc showed the most deterioration when compared with the Voyager data. If the dissipation continues at its current rate, the Liberté arc will be gone within one hundred years. Their observation proved that whatever is holding the ring arcs together is not regenerating them fast enough to sustain them.

### Knowledge Gained

Several stellar occultations with Neptune were observed, but only five showed evidence of a ring or arc system. Scientists working at the European Southern Observatory and the Chilean Cerro Tololo Observatory both witnessed the same brief occultation of star SAO 186001. The two teams noticed that the planet itself did not block the star, because the starlight diminished by only about 35 percent. Neither group was able to locate

*A wide-angle image from Voyager 2, captured in 1989, shows two main Neptunian rings and a faint, inner ring, behind which stars shine. The long exposure resulted in the brightness of Neptune's disk. (NASA)*

any reduction of light on the opposite side of Neptune. These data led to the conclusion that Neptune had only a partial ring or arc system.

The Voyager program consisted of two spacecraft focused on studying the outer planets of our solar system. Launched in 1977, Voyager 2 reached Neptune twelve years later. The spacecraft had two video cameras, infrared and ultraviolet spectrometers, and other instruments. It was not until Voyager 2 started its approach toward Neptune that the first photographs of the rings and arcs were taken. Scientists learned that Neptune in fact did have a mostly complete ring system, like the rest of the gas giants, but that the dust particles that composed the majority of the rings were too small to be detected by Earth-based telescopes. Full rings could be seen only when backlit by the Sun—a view scientists can get only with the use of space probes such as Voyager.

Scientists have more recently been able to use the Hubble Space Telescope's Near Infrared Camera and Multi-Object Spectrometer (NICMOS) to study Neptune's ring system. The camera was powerful enough to detect ring arcs during two different occultations in 1998. One of the difficulties in studying them from Earth is the visual proximity of the rings to Neptune itself. NICMOS solved this problem by using a special filter that blocks wavelengths at which methane, a main component of Neptune's atmosphere, reflects light.

Starting around the year 2000, astronomers were able to use Earth-based telescopes to study Neptune's ring system. Imke de Pater and his colleagues viewed Neptune's ring arcs using the 10-meter Keck Telescope in Hawaii. Advances in image resolution and light-gathering power made this possible. However, the Keck can still detect only the brighter Adams ring; the others remain too faint for it to observe. When the Keck data were analyzed, scientists found evidence that the ring arcs are fading away, with the Liberté arc showing the most damage. If the arcs continue to degrade at the rate they have since 1989, within a hundred years they will have disappeared.

## Context

All of the gas giants in our solar system have ring systems. They all share similarities and have their differences as well. Neptune's system contains five full rings and five arcs. Until telescope technology advances enough to detect the faint dust particles that compose the rings, only the brightest, outermost ring, the Adams ring, will be able to be photographed from Earth.

Learning more about the ring systems promises to help scientists better understand the origins and evolution of the solar system itself. Scientists are also still debating what is preventing Neptune's arcs from forming complete rings. The two main competing theories are based on the idea of shepherding moons that gravitationally trap the particles in arcs. Further advancements must be made before more can be learned about Neptune's ring system, and sending another spacecraft, with more advanced instruments, will be the best way to do so. However, Voyager reached Neptune only through the "gravity assists" of the other planets, which were aligned in a way that occurs only once every 176 years. Short of waiting another century for the planets to realign, the only way to send a new probe to Neptune (preferably one in the Cassini class) would be to develop nuclear propulsion systems that are still in the research and development phases.

*Jennifer L. Campbell*

## Further Reading

Chaisson, Eric, and Steve McMillan. *Astronomy Today*. 6th ed. New York: Addison-Wesley, 2008. This well-written college-level text, designed for introductory astronomy courses, includes a chapter on Uranus and Neptune that covers the ring systems.

Esposito, Larry. *Planetary Rings*. Cambridge, England: Cambridge University Press, 2006. A synopsis of current knowledge of the outer planets' ring systems. Includes information from the Cassini mission and on Neptune's rings and arcs, as well as ring ages and evolution. Geared toward scientists and college students.

Fraknoi, Andrew, David Morrison, and Sidney Wolff. *Voyages to the Stars and Galaxies*. Belmont, Calif.: Brooks/Cole-Thomson Learning, 2006. An introductory college text that gives students easy-to-understand analogies to help explain complex theories. Includes a CD-ROM featuring InfoTrac software.

Freedman, Roger A., and William J. Kaufmann III. *Universe*. 8th ed. New York: W. H. Freeman, 2008. A thorough and well-written college-level introductory astronomy book. Covers all aspects of Neptune, including its ring systems.

Fridman, Alexei M., and Nikolai N. Gorkavyi. *Physics of Planetary Rings: Celestial Mechanics of Continuous Media*. New York: Springer, 1999. Compares the ring systems of Jupiter, Saturn, Uranus, and Neptune using observational and mathematical data. Ideal for scientists, students, or amateur astronomers wishing to know more about the rings of the outer planets.

Miner, Ellis D., and Randii R. Wessen. *Neptune: The Planet, Rings, and Satellites*. New York: Springer, 2002. Covers Voyager 2, its mission, difficulties, and discoveries. Includes a chapter dedicated to Neptune's ring system. Written for nonscientists.

Miner, Ellis D., Randii R. Wessen, and Jeffrey N. Cuzzi. *Planetary Ring Systems*. New York: Springer Praxis, 2006. Looks at the ring systems of each gas giant. Covers recent research in the field, as well as the many questions that remain unanswered.

Tabak, John. *A Look at Neptune*. London: Franklin Watts, 2003. Discusses Voyager 2's mission to Neptune, focusing on what the spacecraft discovered about the planet and what remains unknown. A good introductory work for general audiences.

# NEPTUNE'S SATELLITES

**Categories:** Natural Planetary Satellites; The Neptunian System; Planets and Planetology

*Neptune's family of thirteen known satellites has challenged astronomers to push the limits of their observational technology and theoretical knowledge. While similar in some respects to the satellite systems of the other three Jovian planets, Neptune's satellites, especially Triton, have proven to have unique characteristics, which allow astronomers to increase their understanding of conditions in the early solar system.*

## Overview

Eight days after Neptune's discovery on September 23, 1846, astronomer John Herschel sent a letter to colleague William Lassell, suggesting that he might turn his telescope toward the new planet to search for satellites. Nine days later, on October 10, Lassell discovered what would become known as Triton, the largest satellite of Neptune and the seventh largest satellite in the solar system, with a diameter of 2,706 kilometers. The satellite's name is credited to French astronomer Camille Flammarion and was used unofficially for decades before being officially adopted. Mythologically, Triton was the son of the Greek sea god Poseidon (Roman name, Neptune).

Triton proved to be puzzling to astronomers; unlike most large satellites, it orbits its planet retrograde, or "backward" (opposite to the direction of Neptune's rotation). Its orbit is nearly circular but is highly inclined to Neptune's equator (23°). These anomalous features led astronomers to hypothesize that Triton was captured by Neptune's gravity during a close approach to the planet, despite the fact that other planetary satellites suggested to be likewise captured, such as Saturn's Phoebe, are much smaller.

Despite several searches, no other satellites were discovered around Neptune until the Dutch-born astronomer Gerard Kuiper found a faint (magnitude 19.5) satellite using the 82-inch telescope at McDonald Observatory in 1949. The name Nereid was suggested, after the fifty sea-nymph daughters of Nereus and Doris, attendants of Poseidon in Greek mythology. As strange as Triton's orbit was, Nereid's was no less peculiar. Although it orbits Neptune in a prograde direction at an average distance some fifteen times greater than that of Triton, its orbit is highly elongated (eccentric), with its most distant orbital point nearly seven times farther than its closest approach

to the planet. Such an orbit is more similar to that of a comet than the orbits of most satellites, so—although at a diameter of 340 kilometers Nereid is larger than most asteroids—some astronomers suggested that it also was captured by Neptune's gravity. Others pointed an accusatory finger at Triton, suggesting that the capture of Triton might have disrupted Nereid's original orbit. The discovery of Nereid thus led to more questions than answers, some of which awaited the flyby of the Voyager 2 spacecraft in late August, 1989.

Nereid was too distant (4.7 million kilometers) to be effectively imaged by Voyager 2, which has proven to be a lingering source of frustration to astronomers. The brightness of the satellite is known to change on both short and long timescales, which astronomers have interpreted as due to the rotation of the satellite (although no definitive period can be established). The changes in brightness may be due to differences in the surface material on different sides of the satellite (as in the case of Saturn's satellite Iapetus) or a nonspherical shape. Interestingly, water ice has been spectroscopically found on the surface, leading to speculations that the satellite's surface might be a mixture of ice and some dark material, similar to the satellites of the other Jovian planets (and dissimilar to known Kuiper Belt objects). In addition, Halimede, the innermost of the outer five irregular satellites, has spectroscopic similarities to Nereid, and their orbits have a 0.41 in 1 probability of collision during the history of the solar system. This has led to the suggestion that Halimede is a splinter of a nonspherical Nereid.

The other three Jovian planets were found to have many satellites, which could be divided into two broad classes: regular moons, usually larger in size with prograde, nearly circular orbits nearer the planet; and more distant, irregular satellites, usually smaller and with greater orbital eccentricities and inclinations. Irregular satellites appear to be captured objects, while regular satellites were considered to be the "original" satellites of the planet, dating from the formation of the planet itself. Since Voyager 2 had discovered three new satellites around Jupiter, four at Saturn, and ten orbiting Uranus, it was not surprising when the Voyager 2 imaging team announced the discovery of a new Neptunian satellite in June, 1989, two months before the probe's closest approach to Neptune. Now named Proteus for a Greek sea god, it is actually larger than Nereid (416 kilometers in diameter), although it is exceedingly difficult to observe from earthbound telescopes because of its close orbit (roughly a third Triton's distance from Neptune).

Three more satellites, all roughly one-third the size of Proteus and orbiting closer to Neptune, now named Larissa, Galatea, and Despina, were discovered in late July, 1989. It was later determined that Larissa had been the cause of a brief dip in brightness seen in a star that Neptune had nearly occulted in May, 1981, an observation that had initially been explained as due to the planet's then undiscovered rings. Thalassa and Naiad, each a fraction of the size of Proteus, were discovered orbiting even closer to Neptune only a few days before Voyager's August 25, 1989, closest approach. The names were selected from mythological characters associated with Poseidon/Neptune, according to the International Astronomical Union's (IAU's) convention for Neptunian satellites and based on the historical precedence of Triton and Nereid. For example, Despina was a daughter of Poseidon and Demeter, and Galatea was one of the Nereids. All six of the new satellites were "regular" in the sense that they had prograde orbits that were nearly circular and, with the exception of Naiad, were well aligned with Neptune's equator.

Following the Voyager 2 flyby, numerous attempts were made to discover additional satellites orbiting farther from Neptune than Nereid, which led to the discovery of five "irregular" satellites in 2002 and 2003: in order out from Nereid, these were named Halimede, Sao, Laomedeia, Psamathe, and Neso, all named after individual Nereids in Greek mythology. Sao and Laomedeia have prograde orbits, the other three satellites orbit retrograde, and all five satellites have significant orbital eccentricities (0.3-0.6). Each is between approximately 40 and 60 kilometers wide, comparable to Naiad or smaller, and are largely presumed to have been captured by Neptune.

## Knowledge Gained

Even though Neptune was found to have both "regular" and "irregular" satellites, significant mysteries remained after their discovery. The five innermost satellites were found to be potato-shaped (not unexpected for small satellites whose self-gravity is small). However, some astronomers felt this was more evidence of violence in the Neptunian system. The large size of Triton and its surprising orbit suggested that it had played an important role in the past and future dynamics of the entire satellite system. In 1966, Thomas McCord calculated these effects and found that Triton is in an unstable orbit that eventually will lead to its destruction in possibly tens of millions of years (or more). When it passes inside Neptune's Roche limit, tidal effects will tear it to shreds. McCord

also suggested that Nereid's peculiar orbit was caused by the effects of Triton's capture and later orbital evolution from an initially parabolic or nearly parabolic orbit to a nearly circular one.

Later researchers continued to use various computer simulations and computational techniques to model the evolution of the Neptunian system, in an attempt to explain both the gross and subtle characteristics of each moon's orbit. Many astronomers presume that Neptune had an original generation of satellites, possibly similar to the satellites of Uranus, which was destroyed in the aftermath of Triton's capture, either through mutual collisions between the satellites or through satellites being gravitationally slingshot out of the Neptunian system. The debris is posited to have formed a disk, some of which was accreted by Triton, increasing its diameter and mass, while some of the material accreted to form the currently observed six regular (inner) satellites. The irregular (outer) satellites are suggested to be survivors from this catastrophic event. An unknown number of other satellites collided with each other or one of the larger satellites, or were slingshot out of the system. In the process, the orbits of some of the satellites that were originally in the retrograde-type orbits (normally seen in irregular satellites) were changed into prograde orbits. The orbits of all thirteen satellites have slowly evolved since then through mutual gravitational effects, and these orbits continue to evolve today.

The detailed images taken of Triton by Voyager 2 strengthened the assumption that it was originally a Kuiper Belt object (similar to Pluto). One of the other important discoveries of the Voyager 2 Neptune flyby was Neptune's five rings, named (in order out from Neptune) Galle, Le Verrier, Lassell, Arago, and Adams. As in the case of the other Jovian planets, astronomers expected the ring system and the inner moons to have gravitational interactions. It is suggested that Galatea affects the particles of the Adams ring, which orbit just beyond the orbit of Galatea, although the exact nature of the gravitational resonance is still being debated. Interestingly, it is the Adams ring that has the famous "arcs" (which led to its nickname, the "sausage ring"). Despina, Thalassa, and Naiad orbit between the Le Verrier and Galle rings. Despina has been suggested to be acting as a shepherding moon.

Astronomers continue to explore the details of the satellites' orbital dynamics through cutting-edge computer simulations and theoretical models, and in so doing test our assumptions about the evolution of the solar system in general and the Jovian planets in particular.

## Context

The Voyager 2 mission led to an explosion of knowledge about the Neptunian system, as it did for the Jovian, Saturnian, and Uranian systems before it. However, given the late discovery of the regular satellites, Voyager 2 was able to target only the outermost four of the eight then-known satellites: Triton, Nereid, Proteus, and Larissa.

Little is known about the composition of the other satellites, although they are presumed to have low densities (0.4-0.8) similar to that of Saturn's small satellites Janus, Epimetheus, and Prometheus. Since the exact orbital interactions between the satellites depends on their densities, further research into those dynamics will allow astronomers to put further constraints on the satellites' densities, and hence compositions, and any future observational evidence gathered on the satellites' compositions will lead to further refinements in models of the Neptunian system's dynamic history. Therefore, although Voyager 2 has certainly been a boon to Neptune researchers, further significant breakthroughs certainly await any future spacecraft visiting Neptune and its thirteen satellites.

*Kristine Larsen*

## Further Reading

Cruikshank, Dale P., ed. *Neptune and Triton*. Tucson: University of Arizona Press, 1995. This thick tome is a collection of the invited papers presented at a 1992 scientific conference summarizing what was known about Neptune and its satellites following two years of analysis of the Voyager 2 flyby data.

Cuk, Matija, and Brett J. Gladman. "Constraints on the Orbital Evolution of Triton." *Astrophysical Journal* 626 (2005): 1113-1116. This technical paper details the results of a computer model of the effects of Triton's capture on preexisting satellites.

Holman, Matthew J., et al. "Discovery of Five Irregular Moons of Neptune." *Nature* 430 (2004): 865-867. A technical summary of the methods used to discover the five outermost satellites of Neptune with earthbound telescopes.

McCord, Thomas B. "Dynamical Evolution of the Neptunian System." *Astronomical Journal* 71, no. 7 (1966): 585-590. The first detailed study of the history and future dynamics of Triton's orbit.

Miner, Ellis D., and Randii R. Wessen. *Neptune: The Planet, Rings, and Satellites*. Chichester: Springer, 2002. This book summarizes most of the important results and papers on the Neptunian system through its publication date. The Voyager 2 mission and spacecraft are described in detail.

Moore, Patrick. *The Planet Neptune: An Historical Survey Before Voyager*. 2d ed. New York: John Wiley and Sons, 1996. A popular historical account of the discovery of Neptune, Triton, and Nereid from a famous amateur astronomer and science writer. The epilogue extends the book through the major discoveries of the Voyager 2 mission.

Zhang, Ke, and Douglas Hamilton. "Orbital Resonances in the Inner Neptunian System. I. The 2:1 Proteus-Larissa Mean-Motion Resonance." *Icarus* 188 (2007): 386-399. A technical review of computer simulations of past gravitational interactions between Proteus, Larissa, Triton, and Neptune, with the goal of explaining the current orbital parameters and estimate the densities of Proteus and Larissa.

_____. "Orbital Resonances in the Inner Neptunian System. II. Resonant History of Proteus, Larissa, Galatea, and Despina." *Icarus* 193 (2008): 267-282. A technical investigation of the orbital histories of these satellites as well as estimates of the satellites' densities (and therefore compositions).

# PLANETARY ORBITS

**Category:** Planets and Planetology

*Planets in the solar system revolve around the Sun in elliptical orbits at speeds that vary with distance from the Sun. Laws that govern these motions were first deduced by Johannes Kepler and later quantified by Sir Isaac Newton.*

## Overview

Planets in the solar system move around the Sun in elliptical orbits. Those whose orbits are closest to the Sun move more rapidly than those that are farther away. These simple, universally accepted observations form the basis of the knowledge of planetary motions. Gravity—the force that causes apples to fall from trees and keeps humans firmly planted on Earth's surface—plays the central role in the mechanics of planetary motions.

A simple experiment illustrates the energy relationships inherent in orbiting bodies. If a person attaches a string to a small rubber ball and the ball is swung around

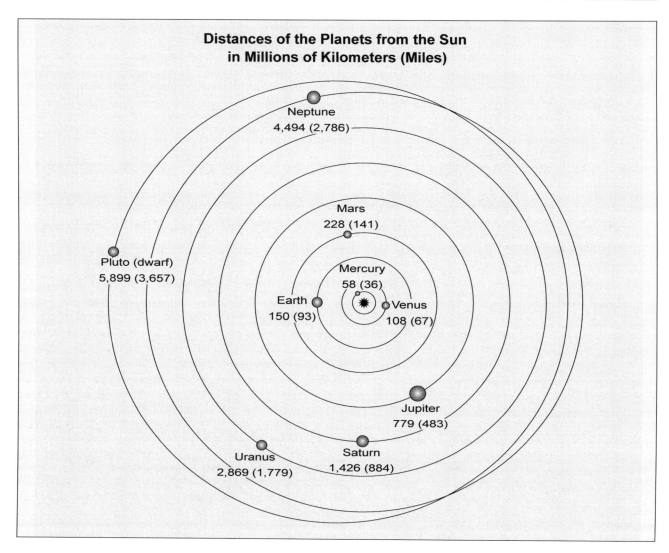

**Distances of the Planets from the Sun
in Millions of Kilometers (Miles)**

Neptune
4,494 (2,786)

Mars
228 (141)

Pluto (dwarf)
5,899 (3,657)

Mercury
58 (36)

Earth
150 (93)

Venus
108 (67)

Jupiter
779 (483)

Uranus
2,869 (1,779)

Saturn
1,426 (884)

the person's head in a horizontal circle, the tension along the string that holds the ball in its "orbit" is analogous to the Sun's gravity pulling on a bound planet. The English astronomer and mathematician Sir Isaac Newton (1642-1727) explained how the force of gravity affects planetary motion. Newton proved in his laws of motion that once an object is in a straight-line motion, it will continue on that course with no further input of energy (law of inertia) unless its motion is perturbed by an unbalanced force. In the case of planets, this force is provided by the gravitational attraction of the Sun (or more massive planet, in the case of a satellite). Depending upon the magnitude of the orbiting body's "kinetic energy" (energy of motion), the body will move in either a circular orbit or, with greater

kinetic energy, an elliptical orbit. Kinetic energy counters the attractive force of gravity, thus preventing a planet from falling into the Sun, or the orbiting ball, as stated in the example, from striking the experimenter.

The scientist who first showed that the orbits of the planets are actually ellipses rather than circles was Johannes Kepler (1571-1630). A German mathematician, astronomer, and astrologer, Kepler worked previously as an assistant to the Danish observational astronomer Tycho Brahe. After Brahe's death, Kepler used detailed position measurements of the planet Mars to plot an orbit that was not circular. Up to this time, planetary orbits—including that of the Moon—were believed to be circular in accordance with precepts developed by the Greek philosopher Aristotle.

A circle is the locus of points all the same distance from a given center. An ellipse differs from a circle in being oval-shaped. An ellipse contains two internal, evenly spaced points called foci. It is important to understand how the foci of an elliptical orbit relate to the positions of an orbiting planet and the Sun. This relationship is expressed by Kepler's first law: Each planet moves around the Sun in an orbit whose shape is that of an ellipse, with the Sun at one focus. The other focus is empty. Thus, the Sun is not precisely in the middle of the ellipse but displaced somewhat to the side. The degree of displacement determines the orbit's eccentricity. As a result, planets move between a minimum distance from the Sun in their orbit, called perihelion, and a maximum distance from the Sun, called aphelion. Planetary orbits have this repetitive pattern due to the central character of gravity; i.e., gravity acts along the line between the gravitationally interacting bodies. The magnitude of the gravitational force follows an inverse square law with regard to its dependence upon distance between the interacting masses. If one doubles the distance between the two objects, their gravitational attraction diminishes not by a factor of two, but by four.

Kepler's second law was actually discovered before his first law. It describes the fact that planets move more slowly when farther away from the Sun (their slowest speed is at aphelion), and they move more rapidly when closer to the Sun (their maximum speed is at perihelion). This observation logically would lend support to the idea that the planet's orbit is anything but circular. The second law states that a straight line joining the planet and the Sun sweeps out equal areas in space in equal intervals of time. Imagine a string attached to a planet at one end and the Sun at the other. When the planet is near aphelion (farthest from the Sun), it moves slowly, so that the triangular sector swept out by the string during a given time will resemble a long, slender piece of pie. In contrast, near perihelion over the same time period, the planet will move farther (because it is going faster), so that the sector swept out by the string resembles a fatter slice of pie. Kepler's second law states that these two pie slices, or triangular sectors—although quite different in radius and opening angle—should have equal areas. This exercise is a mathematical way of stating that planets move more slowly as the Sun-planet distance increases. Planetary orbits obey Kepler's second law of motion as a consequence of conservation of angular momentum.

Kepler's third law, formulated in 1619, was an attempt to quantify the fact that a planet moves more slowly the farther its orbit is from the Sun. His task was to determine a precise mathematical relationship between a planet's average distance from the Sun and its period. Being oval, ellipses have a major axis and a minor axis of different lengths. A line passing through the two foci of the ellipse and ending at both ends of the figure defines the long axis of the ellipse and is known as the major axis. A length equal to one-half the major axis is called the semimajor axis. A line perpendicular to the major axis passing halfway between the two foci of the ellipse is the minor axis. A length equal to one-half the minor axis is called the semiminor axis. A planet's mean distance is half the sum of the perihelion and aphelion distances. This is equal to the average distance of a planet from the Sun and also is the value of the semimajor axis. Kepler found that the cube of the mean distance for any planet is equal to the square of that planet's period. This equation is expressed mathematically as $p^2 = r^3$, where $p$ is the planet's period in Earth years and $r$ is the planet's mean distance from the Sun, expressed in terms relative to the Earth's mean distance, 150 million kilometers, or one astronomical unit (AU). If the Earth's mean distance equals 1.0 AU, then Mars's mean distance is 1.5 AU, Venus's mean distance is 0.72 AU, and so on. Planetary orbits obey this third Keplerian law of motion as a result of the central character of gravity as well as its inverse-square-law nature of gravity.

Newton later reformulated Kepler's three laws using more sophisticated mathematics than was available to Kepler. Newton's modification of the first law states that each planet has an elliptical orbit with the center of mass between it and the Sun at one focus. The "center of mass" is a point between the two bodies (the Sun and the orbiting planet) where their masses are essentially balanced. Mathematically, it is the point at which the product of mass times length is equal for the two bodies: $M_1 L_1 = M_2 L_2$, where $M$ = mass, $L$ = length from center of mass, and subscripts 1 and 2 referring to bodies 1 and 2. The Sun is such an extremely massive body that its center of mass with any planet lies near the Sun's own center. Therefore, the Sun does lie essentially at a focus of the planetary ellipse, as Kepler stated. Its movement around the center of mass (deep within its interior) is detectable only as a slight wobble. For bodies that are more comparable to one another in terms of mass, such as pairs of stars, these objects actually revolve around a common point that lies between them. Pluto and its similarly sized satellite Charon provide a good example of that effect. Because the masses of these bodies are similar, they revolve around a common point known as a the barycenter.

Newton revised Kepler's second law as follows: Angular momentum in a two-body system is constant when no net external torque is present. This law originally described the fact that planets move more rapidly when they are closer to the Sun compared to when they are farther away. Newton found that all bodies that rotate or move around some center have "angular momentum." This quantity is expressed as a body's mass times its speed times its distance from the center of mass ($mvr$, where $m$ = mass, $v$ = linear speed, and $r$ = distance from the center of mass). Because angular momentum is constant for any two-body system in the absence of a net external torque, if $r$ becomes greater, $v$ must become smaller to compensate (mass always remains constant). On the other hand, near the center of mass (the Sun, for planets), the distance $r$ is diminished and speed $v$ must increase to compensate. Conservation of angular momentum comes into play when a spinning skater pulls her outstretched arms close to her body, initiating a more rapid spin rate. Physicists and astronomers usually talk about planetary speeds of revolution or more properly angular velocity, which is the linear speed per unit distance from the focus. In such a discussion, angular momentum then involves the product of moment of inertia times angular velocity. There is no net torque acting on the planet revolving about the Sun, so this angular momentum expression is conserved or remains constant. That means that the distance from the Sun squared times angular velocity is an invariant throughout the planet's orbital motion.

Newton's revision of Kepler's third law is especially important. Newton discovered that the sum of the masses of the two bodies times the square of the period is proportional to the cube of the mean distance, which is expressed mathematically as $(M_1 + M_2)P^2 = a^3$. The masses must be expressed as a fraction of the Sun's mass for the calculation to be valid. The immediate consequence of this equation is that astronomers could now use this equation to calculate the masses of distant bodies given information on the mean distance and period of the orbiting bodies. In most instances, the mass of the smaller body (planet or satellite of a planet) may be neglected because that mass is so insignificant compared to the Sun's mass ($1.99 \times 10^{30}$ kilograms, or 332,943 times Earth's mass). Rearranging the equation gives $M_1 = a^3/p^2$. This equation can now be used to calculate the mass $M_1$ of any central body that has a satellite of mass $M_2$.

## Applications

Consequences of Kepler's and Newton's laws of planetary motion and gravity had an impact on the scientific world not only during their own time but to this very day. The results of their work continue to be used by astronomers to solve problems. For example, the flight path of the Apollo astronauts to the Moon and back was calculated using all three of Kepler's laws. The energy required to propel the Saturn 5 rocket on its way and later to orbit the Moon was calculated using Newton's laws of gravity. The same can be said for all interplanetary spacecraft, such as Voyagers 1 and 2, which visited and photographed the outer planets, Jupiter, Saturn, Uranus, and Neptune. The two Voyager probes were assisted in their journeys by using the gravitational attraction of these massive planets to accelerate them toward their next target. Calculating gravity assists involves kinetic energy and gravitational relationships developed by Newton.

One of the most useful of Kepler's laws for planetary astronomers is the third law as modified by Newton. It allows calculation of the mass of a massive body using data about mean distance and period of one or more of its satellites. It has been used to calculate the masses of all planets that have satellites (which excludes Mercury and Venus). One of the most difficult mass determinations was for the dwarf planet Pluto and its satellite Charon. These bodies are so far from Earth that Charon wasn't discovered until 1977. Its orbital characteristics were determined, with great difficulty, some time later. The similar masses of Charon and Pluto cause them to orbit a center of mass (a barycenter) that lies nearly halfway between them, but the location of that barycenter is somewhat closer to Pluto than it is to Charon. The third law was used to calculate both the mass of Pluto, using data from Charon's orbit, and the mass of Charon, using data for Pluto. These calculations show that both bodies have very low masses and are most likely composed of methane ice.

Another important consequence of the laws of planetary motion involves the survival of life on Earth. One theory suggests that periodic mass-extinction events—such as the demise of the dinosaurs—may have been caused by gigantic impacts of asteroids (rocky planetoids with diameters of less then 1,000 kilometers) or comets (asteroid-sized ice balls) with Earth. In the solar system, most asteroids are concentrated in a belt between Mars and Jupiter, whereas most comets originate in the outer regions of the solar system and beyond. Occasionally, collisions or gravitational perturbations from the massive gas giant planets, such as Jupiter or Saturn, cause asteroids and comets to assume orbits that carry them near Earth. All these bodies have sufficient kinetic energy to resist Earth's gravitational attraction, so that objects that graze Earth's orbit continue by without going into orbit around Earth. This fact explains why Earth and other relatively low-mass planets have few or no satellites (while the gas giants—Jupiter, Saturn, Uranus, and Neptune— have many). Therefore, the bodies that do strike Earth, causing extinctions and making huge craters if they strike land areas, must make a direct hit of a moving Earth. Chances of that occurring on a frequent basis fortunately are rather low, but not zero. Given the billions of years of the history of Earth, it is probable that an occasional body will crash into Earth with catastrophic consequences. The high kinetic energy of these bodies is converted into heat and shock waves upon impact, causing considerable destruction. Newton's laws of gravity and motion play a pivotal role, mostly in determining the trajectories of these dangerous visitors to the inner solar system. By the same token, Newton's laws reveal ways that gravity could be ingeniously used to push possible impacting bodies away from a trajectory that otherwise would have them intersect with Earth, thereby averting a possibly cataclysmic collision.

## Context

The history of science closely parallels the development of astronomy in that the study of heavenly bodies and their relationship to Earth dominated philosophical and religious thinking for millennia. One of the first scientists to study religious thinking and astronomical phenomena seriously was the Greek philosopher Aristotle (384-322 B.C.E.). Unlike most of his contemporaries, Aristotle used some observations to prove his speculations. His major contribution to planetary motion studies was his belief that the natural state of matter is to seek the center of the Earth, which is why objects always fall when released above the Earth. Although erroneous, this and related ideas laid the groundwork for later studies by Galileo and Newton on the effects of gravity. Aristotle also believed, as did many others, that Earth was at the center of the universe. That the Sun and planets revolved around Earth in perfectly

circular orbits was advocated first by his great mentor, Plato. Later, Aristarchus (c. 270 B.C.E.), a Greek astronomer, adopted the idea that the Sun is at the center of the known universe. That idea was forgotten until revived nearly two thousand years later by Nicolaus Copernicus, whose "heliocentric" model, published shortly after his death in a volume titled *De revolutionibus orbium coelestium* (1543; *On the Revolutions of the Heavenly Spheres*, 1952; better known as *De revolutionibus*), describes a system in which the planets orbit the Sun in perfect circles. Although not completely accurate, the heliocentric model eventually supplanted the Earth-centered model of Aristotle and other philosophers.

In the middle of the second century C.E., Ptolemy (c. 100-178 C.E.), wrote a text called *Mathēmatikē syntaxis* (c. 150 C.E.; *Almagest*, 1948) in which he summarized all that was known about astronomy up to that time. This book influenced astronomical thinking for the next millennium. It included a model of the solar system that was quite accurate in predicting planetary positions. Using an idea first developed by Apollonius of Perga (c. 240-170 B.C.E.), Ptolemy declared that planets move in perfect circles around Earth. Nevertheless, planets moved simultaneously in smaller circles called "epicycles." These were necessary to explain why the outer planets occasionally

---

### Kepler's Laws of Planetary Motion

Johannes Kepler's three laws of motion, articulated in the first years of the seventeenth century, laid the foundation for Sir Isaac Newton's law of universal gravitation.

First Law: A planet orbits the Sun in an ellipse, with the Sun at one of the two foci.
Second Law: The line joining the planet to the Sun sweeps out equal areas in equal times as the planet travels around the ellipse.
Third Law: The ratio of the squares of the revolutionary periods for two planets is equal to the ratio of the cubes of their semimajor axes. That is, the time it takes a planet to complete its orbit is proportional to the cube of its average distance from the Sun. The farther from the Sun an object is, the more slowly it moves.

*Johannes Kepler.* (Library of Congress)

seemed to reverse their normally eastern motion relative to the stars and move west. One of the great triumphs of Kepler's laws is that they provide an explanation for retrograde motion (Earth moves faster and overtakes the outer planets). Kepler, and before him Copernicus, laid the foundations for a scientific understanding of planetary motions that broke the hold on thinking imposed by the *Almagest*.

Sir Isaac Newton used ideas developed by Galileo and Kepler to quantify the knowledge of planetary motion and to explain these motions in terms of gravitational forces and kinetic energies. He published his ideas in *Philosophiae Naturalis Principia Mathematica* (1687; *The Mathematical Principles of Natural Philosophy*, 1729; best known as the *Principia*). Although it is now known that Newton's laws do not work well on atomic and subatomic scales (treated in the discipline of quantum mechanics) or in cases where bodies are moving relative to one another at very high speeds close to that of light (which later were addressed by Albert Einstein's relativity theory), Newton's laws work perfectly under everyday conditions on Earth and in the solar system.

Newton's and Kepler's time-tested laws continue to be used by astronomers and other space scientists to make predictions about planetary motions and interactions. Relativity is needed, however, to explain the advance of Mercury's perihelion as it orbits so close to the Sun. Spacecraft deep in the gravitational well of the Sun likewise require calculations involving relativity to maintain them on their proper courses.

*John L. Berkley*

## Further Reading

Arny, Thomas T. *Explorations: An Introduction to Astronomy*. 3d ed. New York: McGraw-Hill, 2003. A general astronomy text for the nonscientist. Includes an interactive CD-ROM and is updated with a Web site.

Beatty, J. Kelly, Carolyn Collins Petersen, and Andrew Chaikin, eds. *The New Solar System*. 4th ed. Cambridge, Mass.: Sky, 1999. Filled with color diagrams and photographs, a popular work on solar-system astronomy and planetary exploration through the Mars Pathfinder and Galileo missions. Accessible to the astronomy enthusiast. Provokes excitement in the general reader, who gains an explanation of the need for greater understanding of the universe around us.

Consolmagno, Guy. *Worlds Apart: A Textbook in Planetary Sciences*. Englewood Cliffs, N.J.: Prentice Hall,
1994. A text accessible to college-level science majors and general readers alike. Presents explanations using low-level mathematics and also involves integral calculus where required. Demonstrates how the area of planetary science progresses by questioning previous understanding in the light of new observations.

Halliday, David, Robert Resnick, and Jearl Walker. *Fundamentals of Physics, Extended*. 9th ed. New York: Wiley, 2007. This textbook has taught millions of college students the fundamentals of physics. Its sections on Newton's laws of motion are particularly strong, as is the chapter on gravitation, which includes derivations of Kepler's laws of planetary motion. Even for those not familiar with basic calculus, there is much to be gained by studying from this all-encompassing work.

Hartmann, William K. *Moons and Planets*. 5th ed. Belmont, Calif.: Thomson Brooks/Cole, 2005. A college-level text that is clearly written; most nonspecialists should find this book a rich source of information. Chapter 3, "Celestial Mechanics," discusses the historical development and application of the laws of gravity and motion. Contains detailed black-and-white diagrams and photographs. Tables in the appendix offer comprehensive data on planetary orbital characteristics and other useful information.

Karttunen, H. P., et al., eds. *Fundamental Astronomy*. 5th ed. New York: Springer, 2007. A well-used university textbook in introductory astronomy. Contains some calculus-based treatments for those who find the standard treatise for typical ASTRO 101 classes too low level. Suitable for an audience with varied science and mathematical backgrounds. Covers all topics from solar-system objects to cosmology.

Leverington, David. *Babylon to Voyager and Beyond: A History of Planetary Astronomy*. New York: Cambridge University Press, 2003. Takes a historical approach to planetary science. Heavily illustrated, concluding with a summary of spacecraft discoveries. Suitable for general readers and the astronomy community alike.

McBride, Neil, and Iain Gilmour, eds. *An Introduction to the Solar System*. Cambridge, England: Cambridge University Press, 2004. A complete description of solar-system astronomy suitable for an introductory college course as well as nonscientists. Filled with supplemental learning aids and solved student exercises. A Web site is available for educator support.

Serway, Raymond A., Jerry S. Faughn, Chris Vullie, and Charles A. Bennet. *College Physics*. 7th ed. New

York: Brooks/Cole, 2005. A textbook used at the introductory level in college physics courses, filled with sample problems, including those on laws of motion. Comes with an online teacher/student resource.

Snow, Theodore P. *The Dynamic Universe*. Rev. ed. St. Paul, Minn.: West, 1991. A general introductory text on astronomy. Covers the kinematics and dynamics of planetary motion. Features special inserts, guest editorials, and a list of additional readings at the end of each chapter. College level.

Stephenson, Bruce. *Kepler's Physical Astronomy*. Princeton, N.J.: Springer, 1994. A complete historical account of the search to understand the orbital behavior of planets in the solar system. For both the technical and general reader.

# PLANETARY RING SYSTEMS

**Category:** Planets and Planetology

*A planetary ring system consists of enormous numbers of relatively small particles that fan out from a planet in the form of a disk, orbiting as a complex unit around that planet. Planetary rings are common in the outer solar system; each of the four "gas giant" planets has a ring system of different age and degree of complexity.*

## Overview

The rings of Saturn are stunning, perhaps among the most beautiful features to observe in the night sky. They were first discovered in 1610 by Galileo in Padua, Italy. Galileo's telescope was not the best of astronomical instruments. What he sketched were two spheres, one on each side of the planet. Galileo thought he had discovered that Saturn was a triple planet. The matter was clarified in 1655, when Dutch observer and telescope maker Christiaan Huygens clearly saw Saturn's rings through his improved instrument. Later, in 1675, Gian Domenico Cassini, supervisor of the Paris Observatory, discovered that there was structure to the rings. An opening in the ring exists about two-thirds of the way out from the planet. This gap became known as the Cassini division. The outer ring was called the A ring and the inner ring, the B ring.

The rings of Saturn were of intense interest to scientists throughout the eighteenth and nineteenth centuries.

New gaps and subrings were subsequently discovered and named. In 1789, English scientist Sir William Herschel estimated the rings to be no more than 500 kilometers thick. By 1850, astronomers' telescopes could resolve that the rings were largely transparent and that the edge of the planet could be seen through the formation. In 1848, French mathematician Édouard Roche rigorously proved that if a satellite orbited too close to a planet, tidal forces from that planet would tear it apart into small pieces. The planet also would not allow a satellite to form from small pieces inside this distance. This limit is known as the "Roche limit." The rings of Saturn fell inside the Roche limit. Roche boldly suggested that Saturn had captured a small satellite that had then been broken into billions of tiny pieces by the gas giant's gravitational forces.

In 1857, the University of Cambridge offered a prize to settle the question of whether the rings were rigid, fluid, or made up of small pieces of matter "not mutually coherent." Scottish physicist James Clerk Maxwell presented a mathematical argument that won the prize. In his proof, Maxwell demonstrated that any solid ring would be torn apart by gravitational forces. He also demonstrated that the rings could not be a liquid. Thus, the rings of Saturn were composed of countless individual particles, each in its own independent orbit around the planet.

From the mid-nineteenth century until 1979, discoveries about Saturn's ring system were largely limited to finding new divisions, and it was postulated that the rings were no more than 15 kilometers thick. In 1977, rings around the planet Uranus were observed using a stellar occultation method. In 1979, the National Aeronautics and Space Administration's (NASA's) twin Voyager spacecraft discovered rings around Jupiter. Voyager 2 confirmed Uranus's rings and, in 1989, discovered diminutive ring arcs around Neptune.

Much about all the planetary ring systems can be learned from the study of Saturn's rings, although the composition of each of the giant planets' rings has not been determined as completely as has Saturn's. The Saturnian ring system is composed almost completely of water ice fragments ranging microscopic particles to boulder-sized bodies, whereas the composition of the rings of Jupiter, Uranus, and Neptune is not fully known. Saturn's rings begin only 32,000 kilometers above the planet's clouds and extend outward to 230,000 kilometers. The rings are no more than a few hundred meters thick. There are seven major rings separated by what are still known as "gaps" (including Cassini's gap between the A and B rings). A closer inspection by Voyager revealed that the major rings consist

*Saturn's ring system was once thought to be the only one in the solar system; it is now known to be only the most pronounced of the ring systems of the four Jovian planets.* (NASA/JPL)

of tens of thousands of ringlets that resemble grooves on a vinyl phonograph record. The ringlets are not cleanly separated from one another but appear to exhibit the property of wave propagation through the structure. There are spiral density waves as well as bending waves. Each ringlet differs in width from 100 kilometers to less than 1 kilometer.

The rings are composed almost exclusively of ice fragments whose diameters range from submicron-size to 10 meters or more. Denser regions of the rings appear to be composed largely of smaller particles, while in the gaps, the larger, meter-sized particles dominate. There may be many ring particles as large as 50 kilometers across within the ring system, although the extent of these particles is not known. As of 2009, the Cassini spacecraft continued to provide images from its orbit about Saturn that reveal more structure to the planet's complex ring system.

One of the most astonishing discoveries made by the Voyagers was the tremendous dynamic complexity of the ring system. An unexpected dynamic effect found by Voyager was the spokelike effect seen in movies assembled out of individual sequences of images taken by the spacecraft. Radial spokes extend outward like spokes on a bicycle wheel. They rotate with the rings. Spokes tended to form, widen, then disappear after about an hour. The most widely accepted theory is that spokes are formed by electrostatic forces between the submicron-size particles and the planet. They are short-lived because the orbital period of the inner particles is faster than that of the outer particles of the spoke.

Another of the Voyager discoveries was the existence of what become known as "shepherd satellites." These are tiny satellites (not readily visible from Earth-based telescopes) that orbit on the outside fringes of the rings within some of the gaps. The minuscule gravitational field of these tiny satellites is enough to push ring particles back into the rings. They help define the rings' outer edges (hence the name "shepherd"). Shepherding satellites also exhibit some rather interesting effects on the rings. In the case of the F ring, two shepherding satellites (1980S26 and 1980S27, later named Pandora and Prometheus, respectively) confine the ring. The F ring appears to be discontinuous in places, is intertwined in others, and is knotted and lumpy in others. The current theory is that the F ring's complexity is caused by the slight eccentricity in the orbits of the two shepherding satellites. It is speculated that over time, variant gravitational interactions cause the structural convolutions. Neptune's ring system of partial arcs exhibited similar discontinuities and irregularities, but there was not enough resolution or coverage to discern shepherding satellites.

Even before the Voyager encounter and the discovery of the superior structural details, a concept known as satellite "ring resonance" was advanced to explain what could be observed from Earth-based telescopes. This concept indicates that the shape of the rings and the location of the gaps are determined not only by the orbits of individual ring particles but also by gravitational influences of Saturn and its major satellites. This theory postulates that Mimas and Enceladus have a particular influence on particles of the rings so that the Cassini division is created by the gravitational interaction

of the satellites on the particles in the ring. The discoveries of Voyager supported part of the resonance theory, but shepherding satellites and other gaps that had no resonance explanation in Saturn's system and that of the other giant planets left some questions about its ultimate

---

### Gian Domenico Cassini, Planetary Explorer

In July, 1664, a professor of astronomy at the University of Bologna, Gian Domenico Cassini, made his first major observation: Jupiter was not a perfect sphere but instead was flattened at its poles. Over the next few years he measured Jupiter's rotational period, observed its moons, and discovered discrepancies in his own measurements that, at first, he attributed to light having a finite speed. (However, he later appears to have rejected his own idea, and in 1676 Danish astronomer Ole Rømer would use Cassini's measurements to calculate the speed of light.)

*(Library of Congress)*

In 1666, Cassini observed surface features on Mars, including Syrtis Major. Again, he measured the planet's rotational period using these features and produced a value of 24 hours and 40 minutes—within 3 minutes of the period now accepted. He also attempted to determine the rotational period of Venus, which he calculated as 23 hours, 20 minutes. What Cassini observed to produce this conclusion is unclear; Venus is entirely covered by bright clouds, and its rotational period (about 243 days) became clear only with the advent of radar.

By this time Cassini's measurements had made him famous throughout Europe, and he came to the attention of Jean-Bapiste Colbert, the French minister of finance. At Colbert's suggestion, King Louis XIV invited Cassini to head the new Paris observatory in 1669. In Paris, Cassini discovered two moons of Saturn, Iapetus and Rhea. In 1675, he recognized that Saturn's ring was divided, separated by a dark gap now called the Cassini division. In 1677, Cassini demonstrated that Saturn was flattened at its poles, and in 1684 he discovered two more moons, Dione and Thetys. In 1705, he correctly suggested that Saturn's ring might not be a solid disk but rather a swarm of small objects orbiting the planet. Cassini also observed several comets between 1672 and 1707, as well as Jupiter's Great Red Spot.

During his years at the Paris observatory, Cassini organized a renowned group of astronomers—including Christiaan Huygens, Ole Rømer, and others—known as the Paris School. Trained in engineering, he published several works on flood control, served as inspector of water and waterways, and became superintendent of the Fort Urban fortifications. By 1711 he was blind, and he died in Paris on September 14, 1712. His son Jacques (1677- 1756) became head of the Paris observatory, and his grandson César-François Cassini de Thury (1714-1784) and great-grandson Jacques-Dominique de Cassini (1748-1845) also became noted astronomers.

effect. Those questions were left for the Cassini orbiter to investigate.

The rings of Saturn are flattened along the equatorial belt because of countless energy-dissipating collisions between particles that have, over the millennia, all but eliminated vertical displacements. Such collisions do not affect the circular orbital motion of the particles; hence, the net effect is a disklike flattening.

The ring system at Jupiter was found to be faint and largely made of dust. Between Voyager and Galileo spacecraft observations, it has been determined that Jupiter's ring system has four components. There is an inner Halo Ring, a relatively thick torus-shaped collection of particles. The Main Ring is very thin and relatively bright. Then there are two Gossamer Rings that are rather wide, thick, and very faint. One is associated with the satellite Amalthea, and the other with the satellite Thebe; these Gossamer Rings are believed to be composed of material coming from their associated satellites.

Astronomers had not expected Jupiter to have rings. Before Voyager, rings there had never been observed from Earth. As Voyager passed behind Jupiter on its outbound journey, it looked back on the giant planet and photographed the ring particles in reflected sunlight. It found that the Jovian rings absorbed all but one ten-thousandth of the sunlight incident upon them. There are two main rings around Jupiter, at 47,000 and 53,000 kilometers above the cloud tops. One ring is 5,000 kilometers wide, while the outer ring is only 800 kilometers in width. There is a torus of thick particles that is not particularly bright, and the small satellites Amalthea and Thebe produce the particles that compose the two Gossamer Rings. Voyager was unable to detect ring thicknesses precisely, but they were estimated to be between 1 and 30 kilometers wide. Voyager scientists estimated that most of the Jovian ring particles are probably made up of dust-size pieces, each in an individual orbit around Jupiter with an orbital period from five to seven hours. Because the tiny, dark particles are in unstable orbits and are constantly falling in toward Jupiter, the rings are probably constantly renewed by the tiny satellite Adrastea, which was also discovered by Voyager.

The rings of Uranus were actually detected before the Voyager spacecraft arrived at that planet. Voyager discovered five well-defined rings around Uranus and four other, less defined rings. The innermost Uranian ring appeared quite compact and dense, while the outermost ring was quite diffuse. Aside from the nine distinct rings around Uranus, from fifty to one hundred nebulous dust bands were discovered in outer orbits containing very small particles. It is speculated that the Uranian ring system began when ten to twelve small satellites less than 200 kilometers in diameter began to break up. As they broke up in low orbit, the fragments began to collide with one another, forming the dust bands and main ring system. Thus, the Uranian ring system may be constantly replenished so that it may have a lifetime of millions of years.

Voyager 2 flew past Neptune in August, 1989, and discovered that this gas giant planet also has a distinct ring system. Voyager found five incomplete ring arcs around Neptune. The exceptionally thin, bright, innermost ring of Neptune was found to be denser than the others, some 17 kilometers wide with a distinct, more diffuse component extending 50 kilometers in toward Neptune. In the outermost rings, distinct points of light were observed, suggesting that the rings are embedded with tiny, icy moonlets, which may act as shepherding agents, as in the Saturnian system. These tiny moonlets are estimated to be about 10 or 15 kilometers in diameter. The outer rings of Neptune are much wider than the inner rings; the third ring is quite dispersed and tenuous but is some 2,500 kilometers across.

The rings of all the giant planets may have been formed by one of several mechanisms. Ring particles may have been accumulated from the breakup of a satellite destroyed by tidal forces; they may have been created when a satellite and asteroid or comet collided in orbit; or the rings may be merely particles left over from the formation of the planet inside the Roche limit that could not accrete into a satellite because of the tidal forces. Why the rings of Saturn are so different from those of the other giant planets may be explained by different material origins. It is possible that the rings around Saturn were created when a satellite of icy origin broke up, while the dark rings of Jupiter represent the remains of a body made of much darker material. None of these conjectures will be proved until pieces of the rings can be directly analyzed.

## Methods of Study

The rings around Saturn are the best understood of the planetary ring systems because their existence was well known long before the Voyagers were launched. Much of the information gained at Saturn by the Voyagers and later the Cassini orbiter can be applied to the other giant planet ring systems, even though the other planets' rings appear much different.

As an orbital platform capable of examining the complex ring system around Saturn with sensors viewing in

*Saturn's B ring, as seen from the Cassini spacecraft in 2008.* (NASA/JPL/Space Science Institute)

several portions of the electromagnetic spectrum in addition to the visible, the Cassini spacecraft verified a number of the Voyager findings and provided even newer insights into ring dynamics, the spokes phenomenon, and gravitational interactions between ring particles and shepherding satellites. Cassini arrived in Saturn orbit on July 1, 2004, and finished a four-year primary mission. Fortunately, Cassini was given an extended mission in light of the spacecraft's good health and constant stream of new and exciting data as well as ongoing stunning photography. In addition to all the refinements of Saturn's ring structure, Cassini produced hints of a tenuous ring system around Saturn's second-largest satellite, Rhea. Cassini used its Magnetospheric Imaging Instrument (MIMI) to infer these thin rings based on a depletion of energetic electrons near that satellite. MIMI data indicated three drops in electron concentration symmetrically located about Rhea, suggesting that particles from decimeters to meters in size were absorbing the electrons. The rings were too tenuous and dark to image directly.

The evolution of planetary ring systems may all be the same. A careful analysis of ring system material should eventually demonstrate this fact. The dynamics of ring systems seem to be relatively similar. Braided and discontinuous rings were seen in all ring systems. Shepherding satellites have been discovered in at least two of the systems. Spokes seemed to be confined to Saturn, which may have something to do with the fact that the mass of material in the Saturnian system was significantly higher than in any of the others.

The question of a ring system's total mass is an important one and awaits resolution. It may be that ring systems have definite lives. If there is no shepherding action, many of the ring particles may be spun out of the ring system by multiple piece encounters, along with random gravitational and tidal effects from the planet and its larger satellites. It may be that the complex Saturnian ring system is relatively young when compared with the other planets' rings; it appears to be better developed. A robust, more massive ring system, thus, may be a sign only of its youth, while a devitalized ring system such as seen on the other gas giant planets may be evidence of their advanced age. This hypothesis, however, remains speculative at best.

A study of ring systems has direct applications to the study of newly forming planetary systems. In this comparison, Saturn can be visualized as a forming star, while the ring particles can be seen as the developing stellar nebula of dust and particles. Such a system has been theorized for our solar system's development some 4.5 billion years ago. In this comparison, the behavior and dynamics of the particles in planetary ring systems can be compared with the early solar system. Even though the Saturnian ring particles are prevented from accretion by confinement within the Roche limit, some comparisons may be made with other dynamic system components.

Such a comparison is not only valuable for direct association to the solar system but also valuable elsewhere in the galaxy. A careful study of the dynamics of planetary ring systems may ultimately be used to calculate the probability of other nebular systems developing around other solar systems. An extension of such estimates will make possible more accurate calculation of the number of planetary systems, where extrasolar planets develop with respect to their star, and perhaps even how many planets could be expected to develop.

## Context

The Saturnian rings have long held a special place in the history of science. Because of their spectacular beauty, they have always been considered to be the crown jewel of the solar system. The application of science to the study of those rings began with sketches by the first scientist to view them, the reputed inventor of the telescope, Galileo. The rings became the focal point for one of the first evaluations of the nature of the solar system from Earth. Mathematical evaluation of the rings accomplished by Maxwell foretold of an era of scientific investigation without direct visitation.

The Pioneer, Voyager, Galileo, and Cassini spacecraft extended humanity's reach to the ringed planets and greatly enhanced the methods of remote study. Evaluating the mass of data returned by robotic probes will continue to keep theorists busy for years to come. Enigmatic spokes, braided rings, and discontinuities each present problems that are still far from being solved. Ring resonance and the concept of ringlets as manifestations of a kind of particulate periodicity also have recently been modeled. Some of remaining dynamic and compositional questions will almost certainly require direct return of samples and other visits by robotic probes. Explorations of the rings will undoubtedly continue, and will provide astronomers and planetary scientists with exciting scientific opportunities for decades.

*Dennis Chamberland*

## Further Reading

Beatty, J. Kelly, Carolyn Collins Petersen, and Andrew Chaikin, eds. *The New Solar System*. 4th ed. Cambridge, Mass.: Sky, 1999. A popular work on solar system astronomy and planetary exploration, filled with color diagrams and photographs. Covers discoveries through the Mars Pathfinder and Galileo missions. Provokes excitement in the general reader, who gains an explanation of the need for greater understanding of the universe.

Bortolotti, Dan. *Exploring Saturn*. New York: Firefly Books, 2003. Full of charts, photographs, a section on observing Saturn, and a historical development of an understanding of the Saturn system from antiquity to the launch of Cassini. For younger readers.

Bredeson, Carmen. *NASA Planetary Spacecraft: Galileo, Magellan, Pathfinder, and Voyager*. New York: Enslow, 2000. This book is part of Enslow's Countdown to Space series and provides an overview of NASA planetary exploration during the last two decades of the twentieth century. Designed for younger readers but suitable for all audiences.

Encrenaz, Thérèse, et al. *The Solar System*. New York: Springer, 2004. A thorough exploration of the solar system from early telescopic observations through the space missions that had investigated all planets with the exception of Pluto by the publication date. Takes an astrophysical approach to place our solar system in a broader context as just one member of similar systems throughout the universe.

Harland, David M. *Mission to Saturn: Cassini and the Huygens Probe*. New York: Springer Praxis, 2002. A technical description of the Cassini program, including its science goals and the instruments used to accomplish those goals. Written before Cassini arrived at Saturn. Provides a historical review of pre-Cassini knowledge of the Saturn system.

Hartmann, William K. *Moons and Planets*. 5th ed. Belmont, Calif.: Thomson Brooks/Cole, 2005. An updated version of a classic text that covers all aspects of planetary science. The material on Jupiter, Saturn, Uranus, and Neptune describes all four ring systems and discusses spacecraft exploration of them. Takes a comparative planetology approach rather than presenting individual chapters on each major object in the solar system.

Irwin, Patrick G. J. *Giant Planets of Our Solar System: An Introduction*. 2d ed. New York: Springer, 2006. Suitable as a textbook for upper-level college courses in planetary science. Focuses on Jupiter, Saturn, Uranus, and Neptune and their satellites, rings, and magnetic fields. Filled with figures and photographs. Accessible to the serious general audience.

Karttunen, H. P., et al., eds. *Fundamental Astronomy*. 5th ed. New York: Springer, 2007. A well-used university textbook in introductory astronomy. Contains some calculus-based treatments for those who find the standard text for introductory astronomy courses too basic. Covers all topics from solar-system objects to cosmology. Suitable for an audience with varied science and mathematical backgrounds.

Miner, Ellis D., Randii R. Wessen, and Jeffrey N. Cuzzi. *Planetary Ring Systems*. New York: Springer Praxis, 2006. Perhaps the most comprehensive text on planetary ring systems that is also accessible to the scientifically inclined reader. Provides interpretation of Pioneer, Voyager, Galileo, and Cassini data and observations. Extensive notes, tables, figures, and references.

Morrison, David. *Voyages to Saturn*. NASA SP-451. Washington, D.C.: Government Printing Office, 1982. This easy-to-read classic NASA publication examines the Voyager encounters with Saturn. Includes details of Voyager's Jupiter encounter and the Jupiter rings. Outlines the history of rings and ring theory. Includes photographs and an index.

# PLANETOLOGY: COMPARATIVE

**Category:** Planets and Planetology

*Spacecraft have obtained detailed photographic, magnetic, radar, and chemical data from the planets Mercury, Venus, Mars, Jupiter, Saturn, Uranus, and Neptune as well as from numerous natural satellites, and even from some asteroids and comets. Data have been used in the preparation of models describing the structure and geological history of planetary and minor bodies throughout the solar system.*

### Overview

Comparative planetology is the study of the broad physical and chemical processes that operate in and on planets over time. It looks for patterns in the similarities and differences displayed by the planets and seeks to provide explanations for them in terms of planetary origins and evolution.

The first successful step in planetary exploration using robotic spacecraft was taken on August 26, 1962, when the National Aeronautics and Space Administration's (NASA's) Mariner 2 spacecraft was launched on a flyby mission to Venus. Mariner 4 was sent to Mars on November 28, 1964. Mariner, Pioneer, Pioneer Venus, Venera, Viking, Voyager, Magellan, Galileo, and Pathfinder space probes have sent back information about Mercury, Venus, Mars, Jupiter, Saturn, Uranus, and Neptune. Long after their primary missions had been completed, some of these spacecraft continued to transmit valuable information back to astronomers on Earth.

The planet Mercury had eluded detailed analysis by astronomers for centuries because of its small size and close proximity to the Sun. Mariner 10, launched in 1973 with a dual mission of studying the clouds of Venus and of photographing Mercury, made three passes by Mercury and was able to photograph about 45 percent of the surface of the planet. Mariner 10 was thus the first spacecraft to take scientific equipment to Mercury. Photographs of Mercury revealed a heavily cratered surface very much like that of Earth's moon. Naturally there are differences between the Moon and Mercury. Since the number of craters per square kilometer varies by as much as a factor of ten, it is believed that some craters may have been covered by a volcanic process. Still, Mercury's surface shows evidence of less volcanic activity than the Moon's. Mercury's largest impact basin, Caloris, has a diameter of 1,300 kilometers. It is believed to have been formed when a large asteroid struck the planet. Photographs show that the shock wave from this collision penetrated the planet and altered the surface on the opposite side, an example of antipodal focusing of seismic energy by the planet's core. Compression scarps (cliffs), which can be as much as 3 kilometers tall and hundreds of kilometers long, were also found on Mercury. They are younger than the craters and are thought to have formed as a result of some internal process such as the cooling of the core of the planet. A magnetic field with a strength of about 1 percent of Earth's magnetic field and a very diffuse atmosphere containing mostly helium were found. Surface temperature ranges from 90 to 948 kelvins.

Venus has been a difficult planet to study from Earth because of its dense atmosphere. Russian Venera probes found a surface temperature of 748 kelvins and an atmospheric pressure of 95 atmospheres. (One atmosphere is the pressure exerted by Earth's atmosphere at sea level.) Photographs of the Venusian surface revealed some areas with smooth plains, while other areas have a rocky terrain. Radar mapping first by the American Pioneer Venus probes and later, in higher resolution and with fuller coverage, by the Magellan orbiter, shows a surface that is 70 percent gently rolling plains, 20 percent lowlands, and 10 percent highlands. The Ishtar Terra highland area of Venus is larger than the United States and includes the mountain Maxwell Montes, which stands about 11.3 kilometers tall. Alpha Regio and Beta Regio are mountainous regions that may contain shield volcanoes. A Russian Vega probe observed lightning discharges in these regions, which could mean

*The solar system planets, in a composite of images captured by missions from Mariner 10 to Cassini.* (NASA/JPL)

other gases, and reflects 76 percent of the light striking it. No planetary magnetic field was found.

Following the Pioneer Venus probes after more than a decade's hiatus, NASA returned to Venus with the Magellan orbiter. The primary objective of Magellan was to obtain a high-resolution map of nearly the entire globe of Venus using a synthetic aperture radar system. Magellan's mission was extended to permit more site-specific investigations. In all, Magellan produced a map of Venus's surface that in resolution and coverage exceeded any available maps of Earth's surface at that time.

Spacecraft such as Mariner, Viking, Pathfinder, and the Mars Exploration Rovers Spirit and Opportunity found that the surface of Mars contains craters, large plains marked by great sand dune areas, chaotic terrain characterized by irregular ridges and depressions, and many volcanoes. The largest volcano, Olympus Mons, stands 27 kilometers tall, has a base diameter of 600 kilometers, and is 64 kilometers across its summit. Orbiting probes uncovered unmistakable signs of catastrophic floods. Impact craters on Mars are concentrated in the southern hemisphere. The northern hemisphere, which has been smoothed by lava flows, has fewer craters, and their features tend to be sharper, indicating that they may be younger than those found in the southern hemisphere. Many of the craters on Mars show evidence of significant erosion. Mars Exploration Rover studies of rocks in situ revealed evidence of sedimentary processes requiring the presence of water.

Mars's seasonal polar caps, made of carbon dioxide, extend well down into the hemisphere, experiencing winter, but shrink and retreat quickly in early summer. Residual polar caps remain throughout the year, although their size does vary. The southern cap is made of carbon dioxide only. The northern polar cap is larger than the southern cap, has a wider temperature variation, and contains mostly water ice. It may be one of the main

that the volcanoes are still active. Lowland areas have the appearance of a cracked slab of lava or cemented volcanic ash. The rocks in the plains are probably granitic rock or potassium-rich basalt.

At the relatively low altitude of 26 kilometers, the Venusian atmosphere is clear, with the temperature dropping to 583 kelvins and the pressure to 20 atmospheres. A thick cloud layer, which is about 80 percent liquid sulfuric acid in the upper portion, exists from 26 kilometers to 60 kilometers above Venus's surface. A sulfuric acid haze exists from 60 kilometers to 80 kilometers altitude. The overall atmosphere is made up of 96 percent carbon dioxide, 3.4 percent nitrogen, and trace amounts of several storehouses for water on Mars. The Mars Polar Lander was designed to investigate that, but it crashed. In 2008, the Mars Phoenix landed on the northern polar region to continue that search for water; by sampling and analyzing the subsurface material, it provided direct evidence that water ice was present in significant amounts.

Mars's atmosphere is 95.3 percent carbon dioxide, 2.7 percent nitrogen, and 1.6 percent argon, with a total pressure of 0.01 atmosphere. The atmospheric temperature varies from 243 down to 173 kelvins. Sublimation of the polar ice caps causes the pressure to vary about 20 percent from season to season. Fog forms in low areas in the early morning. Clouds have been seen around some of the volcanoes. Winds with speeds of at least 150 kilometers per hour pick up surface dust and cause global dust storms. It can take as long as six months for all the dust to settle out from one of these storms. Several spacecraft in orbit at Mars have had to wait for months until the dust cleared in order for their cameras to resume imaging surface features. Long-lived orbital spacecraft have taken images of certain features at widely spaced intervals, showing evidence of wind erosion having altered the surface. Other features have shown evidence of water slumping of crater walls in recent times; that interpretation remains under consideration, however.

The atmosphere of Jupiter is thought to be about 1,000 kilometers thick, with a gaseous composition of 75 percent hydrogen, 24 percent helium, and 1 percent other gases. Pressure at the base of such an atmosphere would be about 100 atmospheres, and the temperature would be about 813 kelvins. The temperature at the top of the atmosphere is only about 113 kelvins. Colored bands, termed zones and belts, are visible in the atmosphere. Zones are yellow-white and represent high-pressure areas where warm currents are rising. Belts are brown, red, or blue-green and represent low-pressure areas where colder gases are sinking. Colors have their source in the interaction of chemical compounds in the atmosphere with sunlight. Very strong wind currents flow in opposite directions where the belts and zones touch. Bands are stable and have not changed their positions for the past one hundred years.

Jupiter's Great Red Spot has been its most prominent feature for over 350 years. It is about 26,000 kilometers from east to west and 14,000 kilometers from north to south. This large cyclonic storm wanders in an east-west fashion and may be stable enough to last for many more centuries. Voyager investigations were followed by the Galileo orbiter and its atmospheric probe, which entered the atmosphere at a point where it found relatively little water vapor.

Jupiter's magnetic field, which is at least ten times stronger than Earth's, produces a radiation belt that is strong enough to kill a human quickly. The radiation belt almost ruined some transistors in the Pioneer probes. Jupiter's radio emissions come from charged particles

## Comparative Data on the Planets of the Solar System

| Parameter | Mercury | Venus | Earth | Mars | Jupiter |
|---|---|---|---|---|---|
| Mass ($10^{24}$ kg) | 0.3302 | 4.8685 | **5.9742** | 0.64185 | 1,898.6 |
| Volume ($10^{10}$ km$^3$) | 6.083 | 92.843 | **108.321** | 16.318 | 143,128 |
| Equatorial radius (km) | 2,439.72 | 6,051.8 | **6,378.1** | 3,396 | 71,492 |
| Ellipticity (oblateness) | 0.0000 | 0.000 | **0.00335** | 0.00648 | 0.06487 |
| Mean density (kg/m$^3$) | 5,427 | 5,243 | **5,515** | 3,933 | 1,326 |
| Surface gravity (m/s$^2$) | 3.70 | 8.87 | **9.80** | 3.71 | 24.79 |
| Surface temperature (Celsius) | −170 to +390 | +450 to +480 | **−88 to +48** | −128 to +24 | −140 |
| Satellites | 0 | 0 | **1** | 2 | 14 |
| Mean distance from Sun millions of km (miles) | 58 (36) | 108 (67) | **150 (93)** | 228 (141) | 779 (483) |
| Rotational period (hrs)* | 1,407.6 | −5,832.5 | **23.93** | 24.63 | 9.9250 |
| Orbital period | 88 days | 224.7 days | **365.25 days** | 687 days | 11.86 yrs |

## Comparative Data on the Planets of the Solar System (*continued*)

| Parameter | Saturn | Uranus | Neptune | Pluto (*dwarf planet*) |
|---|---|---|---|---|
| Mass ($10^{24}$ kg) | 568.46 | 86.832 | 1,102.43 | 0.0125 |
| Volume ($10^{10}$ km³) | 82.713 | 6,833 | 6,254 | 0.715 |
| Equatorial radius (km) | 60,268 | 25,559 | 24,764 | 1,195 |
| Ellipticity (oblateness) | 0.09796 | 0.02293 | 0.01708 | 0.0000 |
| Mean density (kg/m³) | 687 | 1,270 | 1,638 | 1,750 |
| Surface gravity (m/s²) | 8.96 | 8.69 | 11.00 | 0.58 |
| Surface temperature (Celsius) | −160 | −180 | −200 | −238 |
| Satellites | 11 | 5 | 2 | 1 |
| Mean distance from Sun millions of km (miles) | 1,426 (884) | 2,869 (1,779) | 4,494 (2,786) | 5,899 (3,657) |
| Rotational period (hrs)* | 10.656 | −17.24 | 16.11 | −153.3 |
| Orbital period | 29.46 yrs | 84.01 yrs | 164.80 yrs | 247.70 yrs |

trapped in the magnetic field. Voyager 1 found a ring system whose main ring is 6,000 kilometers wide and 30 kilometers thick. A thin sheet of material extends to the surface of the planet.

Jupiter has sixty known moons. Fourteen were discovered by Earth-based astronomers, while two were found in Voyager 1 photographs. Others were found by Voyager 2, the Hubble Space Telescope, and Galileo spacecraft. Active volcanoes were found on the moon Io, Jupiter's innermost moon. An icy crust on Europa is believed to cover an ocean of liquid water; evidence of crustal movement upon such a water layer was found in 2008. That provided strong evidence for internal heating to drive large-scale movements of the icy crust.

Saturn is the second of the giant planets visited by spacecraft. Its atmospheric structure is much like that proposed for Jupiter, but its composition is more like the Sun's, with only 11 percent helium. Belts and zones seen on Jupiter are also visible on Saturn, but their colors are not as intense. The outer layer is predominantly hydrogen. The temperature at the bottom of this layer is 70 kelvins.

Saturn has the most highly developed ring system in the solar system. The ring system has a width of 153,000 kilometers and a thickness of 2 kilometers. There are nine distinct rings, labeled A through G. (Identification of portions of the planet's ring system retains naming schemes from the early days of telescopic observations, before the full complexity of the rings was seen; as a result, six parts of the overall ring system are identified by capital letters, whereas the rest are given names. Unfortunately, letters and names do not necessarily provide information as to distance from the outer atmosphere; for example, the C and B rings are outside the D ring but inside the A ring, and the F and G rings are outside the A ring.) The rings are very complex, made up of an extremely large number of ringlets, some of which are only about two kilometers wide. Shepherding moons orbit around the edge of some rings, and their gravity functions to maintain the sharp edge on the rings. The B ring shows dark features that resemble spokes in a wheel. The particle size in the rings varies from a few thousandths of a centimeter to about ten meters. The spokes rotate as if solid and appear to be particles electrostatically raised above the ring plane.

Saturn has at least sixty satellites. The Cassini spacecraft provided images of many of those satellites detected after the Voyager era. Saturn's only large satellite, Titan, has received a great deal of attention, since it is able to retain a thick atmosphere of nitrogen, methane, and other hydrocarbons. Its surface is obscured due to the thickness of that atmosphere. For that reason, the Cassini orbiter carried a European Space Agency probe named Huygens that was detached from the main spacecraft in order to land on the surface of Titan. Huygens provided evidence of liquid hydrocarbons at its touchdown site existing at cryogenic temperatures. Cassini then was able to image ancient shorelines and prove the existence of lakes of liquid methane across the surface of Titan.

In January, 1986, Voyager 2 passed by Uranus en route to Neptune. Uranus's rotational axis is tilted 82° from the plane in which it orbits. Its rotational direction is retrograde. Voyager 2 data established the rotational period of the planet to be 17 hours, 14 minutes. The greenish atmosphere of Uranus is unusually free of clouds. The primary components of the atmosphere are hydrogen (84 percent), helium (14 percent), and methane (2 percent). The temperature of the atmosphere where the pressure is 1 atmosphere is about 73 kelvins. Voyager 2 observed a tenuous haze around Uranus's rotational pole. This haze is probably formed by the steady irradiation of the planet's upper atmosphere by solar ultraviolet light. Uranus is a weak emitter of thermal radiation from deep within the atmosphere. The planet has been found to be warmer than thought, which implies a greater transparency of the atmosphere than models had predicted. The magnetic field of Uranus is inclined at a 60° angle to the axis of rotation. (Earth's rotational axis and magnetic field are roughly parallel by comparison.)

Voyager 2 found a large spot on Neptune's southern hemisphere in 1989 similar to Jupiter's Great Red Spot. The probe also confirmed the presence of three thin, faint rings around the planet and a magnetosphere. The atmosphere is cold, about 53 kelvins, and its soft, blue tint comes from the presence of methane in the upper atmosphere. Neptune has thirteen known moons, the largest of which, Triton, is covered with methane and nitrogen ices.

The New Horizons spacecraft was launched in January, 2006, to fly by the Pluto-Charon system and thereby complete the initial reconnaissance of all major systems of the solar system. New Horizons was launched when Pluto was still classified as a planet. Later that same year, a new identification system adopted by the International Astronomical Union (IAU) demoted Pluto to the status of a dwarf planet. In June, 2008, the IAU again redefined Pluto, this time as a plutoid, or plutino. Regardless of whether Pluto is a full-fledged planet or a plutoid, New Horizons will provide the first in-depth investigations and closeup images of Pluto and its nearly similar sized satellite Charon sometime in the second decade of the twenty-first century.

## Knowledge Gained

Spacecraft data concerning the atmospheric composition and structure of individual planets have provided significant insight into the solar system as a whole. Mercury's small size and high temperature made it an unlikely candidate for having any measurable atmosphere, yet Mariner 10 found a tenuous atmosphere on the planet. This condition probably arises from the solar wind that bathes Mercury. Venus has a high surface temperature but significantly more mass than Mercury and has been able to retain its atmosphere effectively. Venus's high temperature prevents the buildup of any significant quantity of water, so that carbon dioxide remains in the atmosphere rather than forming carbonates as it can on water-rich Earth. Mars perhaps once had a much denser atmosphere, with large quantities of liquid water—possibly enough to cover the planet to an average depth of 10 meters. Channels on the surface point to large amounts of flowing liquid. As a result of low temperature and low surface gravity, most of Mars's atmosphere has been lost. Perhaps water ice is still trapped below the surface or in the north polar ice cap. Confirming that was the primary objective of the Mars Phoenix mission in 2008, and early results from the lander strongly suggested that white material just underneath the soil was indeed water ice and neither salts nor dry ice. Mars Phoenix was outfitted with a Thermal Evolution and Gas Analyzer (TEGA). Before the end of 2008, TEGA obtained evidence of the presence of water vapor after heating soil samples that were carefully placed within its ovens by a robotic arm equipped with a scoop.

Since the giant planets Jupiter, Saturn, and Uranus, and Neptune have much more massive cores and are much colder than the four inner planets, they can retain light gases such as hydrogen and helium effectively. Differences exist among these four, however, because the core size differs from planet to planet.

The terrestrial planets, Mercury, Venus, Earth, and Mars, show many similar surface features—craters, volcanoes, and mountains. Only Earth has shown activity of its volcanoes, but discharges of lightning around the volcanic mountains on Venus suggests that they may be active also. Volcanism has also been found on Jupiter's satellite Io, Saturn's satellite Enceladus, and Neptune's satellite Triton.

The giant planets all have ring systems, although each system is different from the other three. Saturn's rings are extremely complex, with small divisions between the rings. Uranus has a set of narrow ribbons separated by large spaces, while Neptune has only partially complete ring arcs. Jupiter has a three-component ring system. The innermost portion is called the Halo Ring. Further out is the Main Ring, and that is followed by the wispy Gossamer Rings. High-resolution images from Galileo at Jupiter, Cassini at Saturn, and Voyager 2 at Uranus and Neptune greatly added to the storehouse of knowledge about diversity in ring system dynamics.

One great hope in the exploration of Mars was that some life-form would be discovered. Experiments performed by the Viking landers provided no definitive results. Many astronomers believe that Mars's environment is much too harsh presently to support life as it would exist on Earth. Any primitive non-Earth-like forms of life might be difficult to detect. Life, primitive or otherwise, may also be possible on Europa, Enceladus, or Titan. Few scientists expect to find organisms on the latter two satellites, but some hold out hope that some degree of organized life-forms may be swimming in Europa's ocean under the satellite's icy crust. Until a Europa lander equipped with a subterranean probe can be sent to this satellite, however, that remains only wishful speculation on the part of exobiologists.

## Context

Fascination with outer space is evident when one examines the popularity of space-based science-fiction books, films, and television programs, and when one keeps track of the number of Internet hits on NASA Web sites during high-profile missions like Mars Exploration Rover landings on the Red Planet or space shuttle flights to refurbish the Hubble Space Telescope. Solar system exploration programs are scientific attempts to satisfy human curiosity about space. One fundamental purpose for planetary exploration is to seek a better understanding of the history and perhaps the origin of the solar system. While current models meet some of the criteria, many questions remain. Examination of planetary atmospheres, magnetic fields, ring systems, satellites, and surfaces allows models to be improved and planetary history to be more accurately recorded. For example, the Jupiter and Saturn systems are large enough for them and their satellites to constitute small-scale solar systems. Study of such smaller systems could reveal significant details about the solar system as a whole.

Humanity also has a desire to know whether life exists in any place other than Earth. Are we alone? Is the vastness of the universe just for us, or is it teeming with life? Chances of detecting life in another star system are remote, even if it does exist. The search on the planets of the solar system is much more easily accomplished. In the late twentieth century, both the United States and the former Soviet Union planned uncrewed missions to Mars that would include orbiters, landers, balloons, surface-roving vehicles, and a round-trip mission to return soil samples to Earth. Many of those ambitious plans were delayed considerably, but early in the new millennium an armada of robotic spacecraft orbited around the Red Planet and a number of landers were on the surface searching for evidence of water.

*Dennis R. Flentge*

## Further Reading

Bagenal, Fran, Timothy E. Dowling, and William B. McKinnon, eds. *Jupiter: The Planet, Satellites, and Magnetosphere*. Cambridge, England: Cambridge University Press, 2004. A comprehensive work about the biggest planet in the solar system, comprising a series of articles by experts in theirs field of study. Excellent repository of photography, diagrams, and figures about the Jupiter system and the various spacecraft missions that have unveiled its secrets.

Beattie, Donald A. *Taking Science to the Moon: Lunar Experiments and the Apollo Program*. Baltimore: Johns Hopkins University Press, 2003. Explains the science gleaned from the Apollo lunar landings, including the Apollo Lunar Surface Science Experiment Packages (ALSEPs) and their results.

Briggs, G. A., and F. W. Taylor. *The Cambridge Photographic Atlas of the Planets*. New York: Cambridge University Press, 1982. A collection of the best photographs taken by space probes from the United States and the Soviet Union. In addition to the captions accompanying the photos, a discussion of the important features of each planet and its satellites is provided.

Greenberg, Richard. *Europa the Ocean Moon: Search for an Alien Biosphere*. New York: Springer, 2005. A complete description of current knowledge of Europa through the post-Galileo spacecraft era. Discusses the astrobiological implications of an ocean underneath Europa's icy crust. Well illustrated and readable by both astronomy enthusiasts and college students.

Grinspoon, David Harry. *Venus Revealed: A New Look Below the Clouds of Our Mysterious Twin Planet*. New York: Basic Books, 1998. A thorough examination of the geology of Venus that incorporates Magellan mapping and other data. Explains the Venusian greenhouse effect. A must for the planetary science enthusiast who wants an integrated approach to science and history. Includes speculation about Venus's past.

Harland, David M. *Cassini at Saturn: Huygens Results.* New York: Springer, 2007. Essentially a complete collection of NASA releases from the start of Cassini flight operations through the majority of Cassini's seventy orbits of its primary mission. Provides a thorough explanation of the entire Cassini program, including the Huygens probe's landing on Saturn's largest satellite. Cassini's primary mission concluded a year after this book was published. Technical but accessible to a wide audience.

_____. *Water and the Search for Life on Mars.* New York: Springer Praxis, 2005. A historical review of telescope and spacecraft observations of the Red Planet up through the Spirit and Opportunity rovers. Covers all aspects of Mars exploration but focuses on the search for water, believed to be the most necessary ingredient for life.

Hartmann, William K. *Moons and Planets.* 5th ed. Belmont, Calif.: Thomson Brooks/Cole, 2005. An updated version of a classic text that covers all aspects of planetary science. Particularly strong in its presentation of Earth-Moon science. Takes a comparative planetology approach rather than providing individual chapters on each planet. Examines atmospheres, magnetospheres, satellites, and interiors of the solar system's planets.

Irwin, Patrick G. J. *Giant Planets of Our Solar System: An Introduction.* 2d ed. New York: Springer, 2006. Suitable as a textbook for upper-level college courses in planetary science. Focuses on Jupiter, Saturn, Uranus, and Neptune and their satellites, rings, and magnetic fields. Filled with figures and photographs.

Lovett, Laura, Joan Harvath, and Jeff Cuzzi. *Saturn: A New View.* New York: Harry N. Abrams, 2006. A coffee-table book with about 150 of the best images returned by the Cassini mission to Saturn. Covers the planet, its many satellites, and the complex ring systems.

Morrison, David, and Tobias Owen. *The Planetary System.* 3d ed. San Francisco: Pearson/Addison-Wesley, 2003. A discussion of data from each of the planets accompanied by a large number of photographs and line drawings. Although intended as a college-level astronomy textbook, it provides good reading for anyone with an interest in the solar system.

Squyres, Steve. *Roving Mars: Spirit, Opportunity, and the Exploration of the Red Planet.* New York: Hyperion, 2006. Written by the principal investigator for the Mars Exploration Rovers Spirit and Opportunity, this fascinating book provides a general audience with a behind-the-scenes look at how robotic missions to the planets are planned, funded, developed, and flown. A personal story of excitement, frustrations, a scientist's life during a mission, the satisfaction of overcoming difficulties, and the ongoing thrills of discovery.

# SATURN'S ATMOSPHERE

**Categories:** Planets and Planetology; The Saturnian System

*Data from Pioneer 11, Voyagers 1 and 2, and the Cassini spacecraft, combined with ground-based observations, show Saturn's atmosphere to be composed largely of hydrogen mixed with helium. Clouds of ammonia ice and other chemical components are sources for the various configurations on the planet's visible surface. This hydrogen-helium envelope likely covers a layer of metallic hydrogen that surrounds Saturn's rocky core.*

## Overview

Archaeologists have found recorded observations of Saturn dating from several hundred years B.C.E., inscribed in cuneiform on kiln-baked bricks. It was not until more than two millennia later that Galileo, using one of his early telescopes, was first able to see both the planet and its rings—though he did not recognize them as such. Original drawings made by Galileo indicate that he interpreted the rings as solid. At times he drew them as if they were open handles on a cup, and at other times as filled-in semicircles connected to the planet. In 1794, Sir William Herschel, carefully compared many weeks' worth of observations of subtle markings to deduce a planetary rotation period of 10 hours, 16 minutes, and 0.4 second. In the twentieth century it is realized that Saturn has a mean density of 690 kilograms per cubic meter, considerably less than that of water. Therefore it was realized that Saturn must be composed mainly of the lightest element, hydrogen. The presence of hydrogen was difficult to detect spectroscopically, but methane and ammonia were both observed in 1932.

Ground-based observations of Saturn have always been fruitful, as well as exciting, and continue to be so. However, astronomers' knowledge of the Saturn system was increased dramatically by four spacecraft encounters with the planet. The first was Pioneer 11, later called Pioneer Saturn, which made its closest approach to Saturn

*Saturn from the Hubble Space Telescope in 2004.* (NASA/ESA/Erich Karkoschka, University of Arizona)

on September 1, 1979. It made many important observations of the entire Saturnian system but was concerned mainly with the nature of the surrounding radiation and particle environments. Images made by the photopolarimeter showed a butterscotch-colored planet with an indistinct pattern of subtly varying belts and zones. In other words, Saturn's atmosphere appeared much more subdued than Jupiter's.

More data were gathered during the first of the two Voyager encounters. Voyager 1 made its closest approach to Saturn on November 12, 1980. A few months later, its sister spacecraft, Voyager 2, made its closest approach on August 26, 1981. It was the outbound trajectory of the Voyager 2 spacecraft that was expected to provide good images of Saturn's southern hemisphere. Unfortunately, soon after closest approach, a problem arose with Voyager 2's scan platform, the movable platform upon which several of the instruments—including the two cameras—were mounted. Thus, most of the images obtained during the departure portion of the flyby show primarily the northern hemisphere.

Continuing analysis of all these observations has led to a reasonably consistent picture of Saturn's atmosphere, although much remains uncertain. One of the most obvious features of Saturn is its shape. It is flattened at its poles more than any of the other planets. Measurements estimated by the International Astronomical Union (IAU) in 1985 for Saturn's polar and equatorial radii were 53,543 and 60,000 kilometers, respectively.

The shape of the planet, in association with measurements of other quantities, such as its gravitational field, composition, and heat production, is an important constraint on models of the planet's interior. The major chemical component of Saturn is hydrogen, but other elements are present as well. In particular, a heavy central core, composed of some sort of ice or rock, is proposed. To be consistent with gravity measurements, this core must contain between 10 and 20 percent of Saturn's mass. This proportion is considerably larger than the 2 or 3 percent that would be expected if Saturn had the same composition as the Sun. This central region may extend to about one-fifth of Saturn's radius.

Outside the core lies the hydrogen-helium atmosphere. In the visible part of the atmosphere, hydrogen is in its normal gaseous form, but deeper inside Saturn it is compressed by the weight of the overlying atmosphere. About halfway between the center and the visible surface, the pressure is so great, around 3 million bars, that hydrogen must change form and behave like a metallic fluid. For comparison, the atmospheric pressure at Earth's surface is about one bar.

It is this metallic region that could generate Saturn's magnetic field. Researchers believe that irregularities in this region produce slight asymmetries in the magnetic field, which, in turn, regulates the radio emissions observed by the Voyager spacecraft. It is for this reason that the observed period of 10 hours, 39 minutes, and 24 seconds (with an estimated error of 7 seconds) of this radiation is thought to represent the rotational period of the deep interior of the Saturn.

After hydrogen, the largest constituent of Saturn is helium, the second lightest element. The mass fraction of helium in Saturn has been estimated from infrared measurements to be $0.06 \pm 0.05$, significantly less than the value of $0.18 \pm 0.04$ found for Jupiter. This difference is thought to be caused by the fact that, at the lower temperatures found on Saturn, some of the helium becomes insoluble in the primarily hydrogen atmosphere. It is thought that it

may form small droplets at some level and "rain out," dissolving again at a deeper, and hence warmer, level. This process would warm the atmosphere somewhat.

The observed average temperature of Saturn, derived from infrared measurements, is about 95 kelvins, compared with an equilibrium temperature of 82 kelvins, the temperature that would be expected if the planet were entirely dependent on solar radiation for heat. The implication is that Saturn radiates $1.78 \pm 0.09$ times as much heat as it receives. Some of this energy is thought to be generated in the helium separation.

Because of its 27° orbital inclination, larger than the 23° inclination of Earth, Saturn has seasons. The period of the Voyager encounters corresponded to early spring in the northern hemisphere. The large heat capacity of Saturn's atmosphere, however, meant that the southern hemisphere, where it was early fall, would still be a few kelvins warmer than the northern hemisphere.

Although hydrogen and helium account for most of the mass of Saturn's atmosphere, they are not responsible for

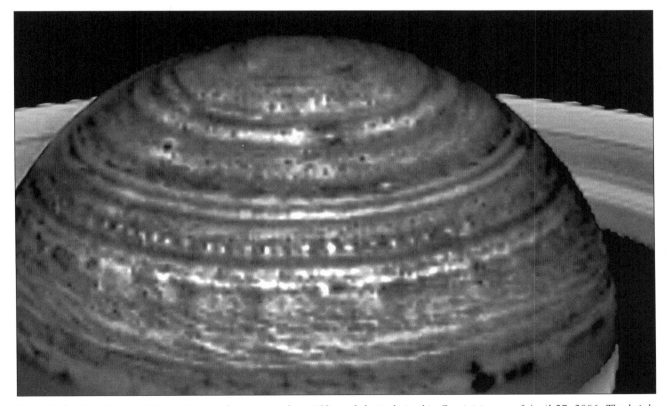

*Saturn's "string of pearls" formation can be seen at about 40° north latitude in this Cassini image of April 27, 2006. The bright "beads" are clearings in the clouds, which reveal the thermal glow of the planet below, and are regularly spaced, suggesting a large planetary wave.* (NASA/JPL/University of Arizona)

the patterns seen in the Voyager images. Despite the fact that only gaseous ammonia has been observed spectroscopically, these patterns are thought to be the tops of ammonia clouds. Theoretical models predict that below the ammonia clouds there are clouds of ammonia hydrosulfide and water ice. There is, however, little observational confirmation of these lower cloud layers. Lack of observational data makes it difficult to model the behavior and possible interactions of these multiple cloud layers. Whether they produce precipitation—ammonia or water precipitation in the form of either rain or snow—remains a matter of speculation.

The presence of ammonia clouds still fails to explain the appearance of the planet, since a cloud of small ammonia droplets would not create Saturn's observed butterscotch color. It should be remembered that the colors of many of the published spacecraft images are considerably enhanced (as "false-color" imagery) to make subtle variations distinguishable. Thus, an additional chemical component, termed a chromophore, has been postulated. Such

a component must be capable both of existing at this level in the atmosphere and of providing the needed color. The complex radiation environment of Saturn's upper atmosphere could produce many types of chemicals. Among those suggested as cloud chromophores are various compounds of sulfur, phosphorus, and hydrazine, as well as various mixtures of organic compounds containing both hydrogen and carbon.

A comparison of two images of Saturn's atmosphere will show not only that there is an interesting variety of cloud shapes but also that they are all moving relative to one another. Most of the movement observed is longitudinal—that is, around the rotational axis. If the radio rotation period is used as the basic rotation rate, then the clouds show an alternating series of east-west jets. Close to the equator, these jets have a velocity of between 400 and 500 meters per second (the so-called equatorial super-rotation). The velocity decreases away from the equator until it reaches a retrograde speed of 25 meters per second at about 40° north and south latitudes. It then increases again to nearly 150 meters per second before decaying yet again. This pattern has been observed to 84° north latitude, always with posigrade jets moving considerably faster than the retrograde ones. The mechanism responsible for generating these jets is still uncertain. The motions in Earth's atmosphere are its response to an uneven solar heat input. The attempt to move warm air poleward produces, in the shallow (less than 1 percent of Earth's radius) rotating atmosphere, streams of high- and low-pressure areas encircling the midlatitudes. A similar mechanism could operate on Saturn, with observed surface motions decaying rapidly with height. An alternative suggestion is that Saturn's internal heat source produces a form of convection, and that the observed zonal flow around the rotational axis extends all the way through the planet, parallel to the rotational axis.

A number of additional features have been noted on Saturn's visible surface. In the southern hemisphere, there is a reddish spot which resembles a miniature version of Jupiter's Great Red Spot. Near the equator, there are various

*A close view of the far-north atmosphere, taken in May, 2008, from the Cassini orbiter, displays swirls and vortices in the hydrogen-helium "ocean." (NASA/JPL/Space Science Institute)*

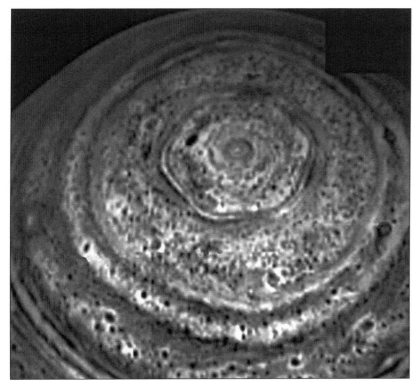

*The hexagonal shape at Saturn's north polar region (about 78° north latitude) was first discovered in the 1980's and seems to persist, appearing in this 2006 image from Cassini.* (NASA/JPL/University of Arizona)

being 5,700 kilometers. This feature apparently represents some type of atmospheric wave. Measurements of the infrared emissions of this region show a large north-south temperature gradient, and it could be that this feature is the same type of atmospheric phenomenon as that which transfers equatorial heat northward on Earth.

Despite the nearly equatorial trajectory of the Voyager spacecraft, some images of the north polar region were obtained. These show it to be populated with many small, fluffy clouds, close to which, at about 80° north latitude, was a regular hexagonal pattern, formed of long, thin, striated clouds that are moving at about 100 meters per second. The hexagonal pattern appeared to be stationary relative to the radio rotation period, with the clouds passing around its corners. The driving mechanism of this feature was a mystery, although the close association between its rotation rate and the radio rotation period could be significant. The Cassini spacecraft returned images in 2007 of Saturn's north polar region that again showed such a hexagonal pattern. This time the hexagon was nearly 24,000 kilometers across. Data suggested that the nearly perfect hexagonal pattern extended down into Saturn's clouds to a depth of nearly 100 kilometers. According to University of Oxford planetary scientist Leigh Fletcher, the appearance of this vortex was surprising. Apparently gas moved toward the pole and was compressed and heated as it dropped into the depths of Saturn's troposphere over the pole. The physical mechanism for this behavior remained unknown.

Cassini entered orbit about Saturn on July 1, 2004. During its primary four-year mission, the spacecraft returned impressive data about the ringed planet and its satellites, including the particles and fields environment surrounding Saturn. The spacecraft remained in nearly perfect health, and therefore it was funded for a two-year extended mission to continue the scientific harvest.

wispy clouds, which appear to be stretched by the strong wind shears present in this region. Farther north, at around 40° north latitude, there are three brown spots, each one a few thousand kilometers across. Just to the south of these is an area in which white blobs of cloud, possibly of convective origin, appear, evolve rapidly, and dissipate. In the mid-1990's, the Hubble Space Telescope detected a large white storm near the equator that eventually grew to nearly 20,000 kilometers in length before beginning to fade. This storm, referred to as Saturn's Great White Spot, is hardly unique. Large white storms of this type have been noted in astronomical records on a nearly thirty-year period going back well into the nineteenth century.

Just to the north of the brown spots, centered at about 47° north latitude, is the "ribbon feature," a fairly light band, about 5° wide, inside of which is a darker streak, around 1,000 kilometers wide, which threads an oscillatory north-south path along the center. The appearance of the individual peaks and troughs evolves rapidly over a few days, the average distance between adjacent peaks

### Knowledge Gained

Spacecraft flybys of the Saturnian system and prolonged orbital observations by Cassini have dramatically increased astronomers' knowledge of Saturn's dynamic and complex atmosphere. Measurements of the hydrogen and helium abundance, the zonal circulation, the structure of the planet's gravitational and magnetic fields, and the derivation of the radio rotation rate would have been very difficult, if not impossible, to obtain from Earth and even from the Hubble Space Telescope.

These Cassini observations have enabled scientists to refine considerably their models of Saturn's internal composition and structure. This model envisages that the visible envelope, composed primarily of hydrogen and helium, extends about halfway to the planet's center; below this covering is a region of metallic hydrogen and a small rocky core. The deep atmosphere is warmed by Saturn's internal heat source and becomes cooler with distance from the core, reaching a minimum of about 85 kelvins at a pressure level of around 100 millibars at the tropopause. Above this level, solar heating becomes significant, and the temperature starts to increase again. Visible clouds are thought to consist of ammonia ice crystals with, possibly, ammonium hydrosulfide and water-ice cloud layers below them. Existence of these lower cloud layers would depend on the amounts of the various elements in this part of the atmosphere.

The Voyagers provided an extensive album of images of Saturn, documenting cloud motions and their morphologies. Such images, however, have increased scientists' understanding of the atmospheric dynamics only to a limited extent. Even the basic driving mechanism remains uncertain; it could be the planet's internal heat source or the Sun's external heat supply. The extent to which Voyager observations have increased scientists' knowledge can also be considered in terms of their understanding of atmospheres in general. The planetary atmosphere most comparable to Saturn's is that of Jupiter, which is slightly larger, rotates slightly faster and has a similar hydrogen and helium composition. Jupiter is, however, much closer to the Sun than Saturn, and it therefore receives several times as much solar energy.

Compared with Jupiter's highly turbulent atmosphere, Saturn's atmosphere, with its subtle belt-zone variations and widely separated spots, might appear to be relatively quiescent. Voyager observations, however, have shown

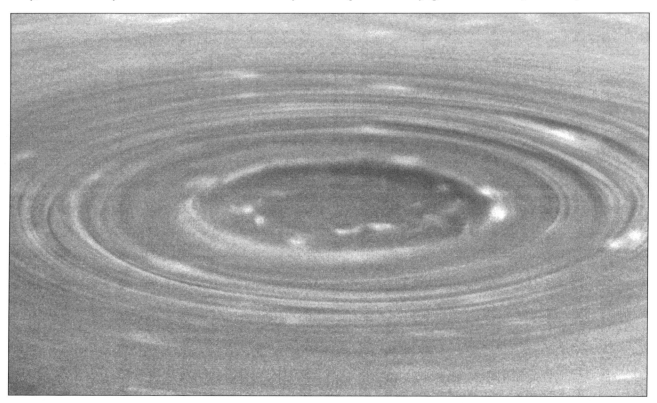

*At Saturn's south polar vortex in this July, 2008, Cassini image, bright clouds can be seen within the inner ring, providing tantalizing clues to how heat energy may move through Saturn's atmosphere. (NASA/JPL/Space Science Institute)*

that this is far from the case. Saturn's atmosphere moves at velocities up to 500 meters per second, compared with about 200 meters per second on Jupiter. Jupiter observations also suggested that Saturn's alternating belt-zone pattern was associated with alternating eastward and westward jets. Saturn observations, which had no such obvious correlation, showed that this model was overly simplistic. Another interesting difference between the two planets is that on Jupiter westward jets reach much the same speed, relative to the radio rotation period, as eastward jets. The situation on Saturn is far less symmetrical, with eastward jets dominating.

Circulation patterns in Earth's atmosphere are still not fully understood. Thus it is not surprising that scientists' knowledge of the atmospheres of the outer planets remains fairly basic. The measurements of the Voyager and Cassini spacecraft contribute to that knowledge mainly by increasing the number of observations upon which models can be based.

One such Cassini observation was referred to by Internet-based National Aeronautics and Space Administration (NASA) reports as Saturn "riding the wave." The wave pattern in the atmosphere is visible from Earth only every fifteen years. Earth has a similar oscillation, but its period is only two years. Jupiter has a similar oscillation, but with a four-year period. Saturn's oscillation in question was under observation with ground-based telescopes and Cassini's Composite Infrared Spectrometer. This oscillation involves temperatures in Saturn's upper atmosphere switching with altitude in a hot-cold pattern that, when graphed in three dimensions, assumes a shape much like the stripes that wrap around a candy cane. These temperature oscillations result in winds changing direction from east to west and back again, and it is this pattern to which scientists were referring as the wave pattern discovered on Saturn.

Beginning in December, 2007, Cassini imaged a storm in Saturn's southern hemisphere that produced lightning discharges with energies more than ten thousand times that of typical lightning on Earth. Cassini's radio and plasma wave instruments actually picked up the lightning about a week before the storm itself could be identified by the spacecraft's imaging cameras. Scientists then used the periodic appearance of the storm to confirm Saturn's rotation rate.

## Context

If it had not been for the Pioneer, Voyager, and Cassini missions, knowledge of Jupiter and Saturn would be limited to that obtainable from Earth or from Earth-orbiting observatories such as the Hubble Space Telescope. Although probes have not entered Saturn's atmosphere, they have still provided many observations that contribute to astronomers' understanding of it. For the first time, researchers could see how the planet changes in appearance when it is viewed from different directions. Monitoring of each spacecraft's radio signal as it disappeared behind Saturn provided information about the local atmospheric temperature profile (the variation of temperature with height). Proximity of these spacecraft to the planet also allowed study of the structure of individual features.

Prior to these dedicated spacecraft missions, Saturn was a greatly appreciated but little understood object. Only its most global properties had been investigated. It was believed to be a typical "gas giant"—a smaller, colder, and less colorful version of Jupiter. Voyager observations revealed this concept to be only partly true. The elemental abundances derived for Saturn are different from those of Jupiter, the most striking difference being the hydrogen-to-helium ratio, which, combined with the planet's unexpectedly high temperature, led to the idea that its helium becomes depleted by becoming insoluble and precipitates into the lower atmosphere.

Very little was known about Saturn's atmospheric dynamics before Voyager images were obtained. Tracking motions of the occasional atmospheric spots visible from Earth had suggested a predominantly zonal flow, strongest near the equator. It was only with the Voyager flybys that periodic radio emissions from the planet were observed. The association of these with rotation of Saturn's interior provided a plausible base velocity against which the motion of other features could be measured. Prior to this finding, velocities could only be given relative to an arbitrary feature.

The Pioneer 11 encounter provided some intriguing information about Saturn's atmosphere; the two Voyager encounters supplied far more. All these flyby missions, however, could provide only a brief look at the planet. Ground-based observations suggest that the large-scale zonal flow is probably fairly stable. The appearance of the individual belts and zones can change, however, while smaller individual features can evolve quite rapidly. Some of these changes may be seasonal effects, while others could be caused by various types of instabilities in the same way that the fairly constant solar heating of Earth produces a perpetual sequence of low- and high-pressure regions.

To investigate these effects further, it was necessary to observe the planet over a longer period than is possible with a flyby mission. The Hubble Space Telescope was deployed from the space shuttle *Discovery* in April, 1990. After a troubled start requiring repair missions, Hubble was able to begin taking unprecedented images, including on occasion some of Saturn. Although Hubble could capture images only at a much lower resolution than the Voyager spacecraft had achieved, Hubble did have the capacity to observe in spectral bands that could not be seen from Earth because of absorption by the atmosphere.

The obvious next step in Saturn investigation was to place an orbiter about the ringed planet, one that could use encounters with its many satellites in addition to propulsion system firings to alter its course so that the spacecraft could swing close to interesting objects; a principal focus on studying the large satellite Titan and its atmosphere was given to this follow-on mission. The National Aeronautics and Space Administration and the European Space Agency named the follow-up probe to the Voyager flyby missions Cassini after the astronomer who discovered several of Saturn's satellites. Cassini was launched on October 15, 1997, and after several gravity assists in the inner solar system, it flew past Jupiter and then was redirected to enter the Saturnian system and conduct its primary science mission. That primary mission ended in 2008, but as the spacecraft was still in fully functional condition, a two-year extended mission was conceived and funded.

Cassini found confounding aspects in Saturn's dynamic and complex atmospheric structure. White spots, which were also seen with the Hubble Space Telescope, were seen to develop and diminish. However, perhaps the most perplexing atmospheric features were a hexagonal pattern about the north pole 25,000 kilometers across and another similar feature found later to exist around the south pole. The nature and stability of features such as these polar ones remains under investigation.

*David Godfrey*

## Further Reading

Alexander, Arthur Francis O'Donel. *The Planet Saturn: A History of Observation, Theory, and Discovery.* New York: Macmillan, 1962. Alexander catalogs the historic observations of Saturn through the ages until approximately the middle of the twentieth century. For anyone interested in the history of Saturnian observations, this is an essential resource. The way the subject matter is divided into very small sections, however, makes some parts of the book difficult to read.

Benton, Julius. *Saturn and How to Observe It.* New York: Springer, 2005. An observer's guide to astronomy focusing on Saturn and what can be seen with modest telescopes. Provides plenty of information about the Saturnian system for the amateur astronomy buff.

Bortolotti, Dan. *Exploring Saturn.* New York: Firefly Books, 2003. A look at the Cassini-Huygens mission for a younger audience. Full of charts, photographs, a section on observing Saturn, and a discussion of the historical development of Saturn studies, from antiquity to the launch of Cassini.

Consolmagno, Guy. *Worlds Apart: A Textbook in Planetary Sciences.* Englewood Cliffs, N.J.: Prentice Hall, 1994. A text aimed at both college science and nonscience majors alike. Presents subjects using low-level mathematics but uses integral calculus where required. Demonstrates how the area of planetary science progresses by questioning previous understandings in light of new observations.

Encrenaz, Thérèse, et al. *The Solar System.* New York: Springer, 2004. A thorough exploration of the solar system, from early telescopic observations through the space missions that had investigated all planets (with the exception of Pluto) by the publication date. Takes an astrophysical approach to place our solar system in a wider context as just one member of similar systems throughout the universe.

European Space Agency. *The Atmospheres of Saturn and Titan.* ESA SP-241. Paris: Author, 1985. This report of the proceedings of a conference held in 1985 covers the then-current understanding of Saturn and its largest moon, Titan, and anticipated the Cassini mission to the two bodies, launched in October, 1997.

Harland, David M. *Cassini at Saturn: Huygens Results.* New York: Springer, 2007. The text provides a thorough explanation of the entire Cassini program, including the Huygens landing on Saturn's largest satellite. Essentially a complete collection of NASA releases from the start of Cassini flight operations through the majority of Cassini's seventy orbits of its primary mission. Cassini's primary mission concluded a year after this book entered print. Technical, but accessible to a wider audience.

_____. *Mission to Saturn: Cassini and the Huygens Probe.* New York: Springer Praxis, 2002. A technical description of the Cassini program, its science goals, and the instruments used to accomplish those goals. Provides a historical review of pre-Cassini knowledge of the Saturn system. Written before Cassini arrived at Saturn.

Hartmann, William K. *Moons and Planets.* 5th ed. Belmont, Calif.: Thomson Brooks/Cole, 2005. An updated version of a classic text that covers all aspects of planetary science. The chapter on Saturn covers the Saturnian system and spacecraft exploration of it.

Hunt, Garry E., and Patrick Moore. *Atlas of Saturn.* London: Mitchell Beazley, 1982. This book, which is accessible to anyone with an interest in the planets, summarizes knowledge of Saturn both before and after the Pioneer and Voyager encounters. Provides detailed descriptions of the Voyager observations, as well as the spacecraft and the instruments that made them.

Irwin, Patrick G. J. *Giant Planets of Our Solar System: An Introduction.* 2d ed. New York: Springer, 2006. Suitable as a textbook for upper-level college courses in planetary science. Focuses on Jupiter, Saturn, Uranus, and Neptune and their satellites, rings, magnetic fields, and atmospheres. Filled with figures and photographs. Accessible to the serious general audience.

Lovett, Laura, Joan Harvath, and Jeff Cuzzi. *Saturn: A New View.* New York: Harry N. Abrams, 2006. A coffee-table book replete with about 150 of the best images returned by the Cassini mission to Saturn. Covers the planet, its many satellites, and the complex ring systems.

Morrison, David. *Voyages to Saturn.* NASA SP-451. Washington, D.C.: Government Printing Office, 1982. Written by a member of the Voyager imaging team, this account of the Voyager missions to the Saturn system describes the people and events that contributed to the encounters and how they changed scientists' understanding of the planet. Includes many of the beautiful color pictures produced from the Voyager imaging data.

Russell, Christopher T. *The Cassini-Huygens Mission: Orbiter Remote Sensing Investigations.* New York: Springer, 2006. Provides a thorough explanation of the remote-sensing investigations of the Cassini orbiter and the Huygens lander. Outlines the scientific objectives of all instruments on the spacecraft and describes the planned forty-four encounters with Saturn's moon Titan. Only the science returns to 2006 are covered.

# SATURN'S INTERIOR

**Categories:** Planets and Planetology; The Saturnian System

*Saturn, the second largest planet in the solar system, is famous for its ring system, but the key to understanding the planet is in understanding its interior structure. Though it is similar to Jupiter and likely shares a similar origin, Saturn is not just a smaller version of Jupiter. While Saturn is called a gas giant planet, its interior is more than just a big ball of gas.*

## Overview

Jupiter and Saturn are the two largest planets in the solar system. Both planets are believed to have formed directly out of the accretion disk of material surrounding the Sun during its formation. As a result, both planets would be expected to have similar compositions and structures. There are important differences between Jupiter and Saturn, however. Because Saturn formed directly from the material accreting to form the Sun, it is expected to have a composition similar to that of the Sun and Jupiter. Indeed, Saturn is composed primarily of hydrogen and helium, the main constituents of Jupiter and the Sun.

Saturn is the second largest planet in the solar system, having about 30 percent of Jupiter's mass, yet that mass is spread over a volume that is about 60 percent that of Jupiter. This gives Saturn a much lower density than that of any other planet in the solar system. One reason for this low density is that hydrogen, the main component of Saturn and Jupiter, is very compressible. Saturn's lower gravity, due to its lower mass, compresses its hydrogen less than does Jupiter. Thus, Saturn is much more distended than Jupiter.

Though Saturn's primary composition is hydrogen and helium, it has also accumulated heavier elements through collisions with smaller bodies since it formed. Some of the material is iron and silicates (rocks), and these materials sink to the center of the planet to form a core. Such a core would be under extreme pressure and temperature, with the temperature of the core is probably near 10,000 kelvins. Since Saturn is in the outer solar system, many of the bodies colliding with it would also contain ices— not only water ice but also frozen carbon dioxide, frozen methane, and frozen ammonia. These ices, heavier than hydrogen, also sink deep into the planet. However, at the pressure and temperature at that depth, the ices are in a liquid state. The term "liquid ices" is often used to describe the state of these fluids.

Saturn has an oblateness of 0.98 (its equatorial diameter is 9.8 percent greater than the pole-to-pole diameter), making Saturn the most oblate of the solar system's planets. Though this level of oblateness constrains the possible size of Saturn's core, there is some disagreement among planetary scientists as to the size of Saturn's core. Many models suggest that Saturn has a core that is about 10 to 15 times the mass of Earth. This is similar in mass to Jupiter's core, but it is less compressed, having a diameter perhaps in excess of 20,000 kilometers. Though Saturn's core is about the size of Jupiter's core, it is a much larger percentage of Saturn than Jupiter's core is of Jupiter. A layer of liquid ices of perhaps up to 10,000 kilometers thick likely sits on top of the core.

The vast majority of the rest of Saturn is composed of hydrogen and helium. These are in a gaseous state in the outermost parts of the planet. Clouds of ammonia ice and other ices exist in the upper few hundred kilometers. Below about 2,000 kilometers beneath the clouds, the pressure is so high that hydrogen is compressed into a liquid state. However, the temperature and pressure of Saturn's interior are well beyond the critical point of hydrogen. The critical point, in chemistry, is the temperature and pressure at which there is no clear distinction between liquid and gas fluid states. Therefore, there is no clear boundary between the gaseous upper portion of Saturn and the liquid hydrogen inner portion. As depth increases, the hydrogen gradually becomes more and more liquid-like. The bulk of Saturn is in this liquid state, even though it is called a "gas" giant planet.

Hydrogen begins to exhibit typical metallic properties, such as electrical conductivity, when it is under sufficient pressure and temperature. On Saturn, these conditions are believed to be met near 25,000 kilometers beneath the cloud layers. Thus, from the top of the liquid ices layer out to nearly 55 percent of Saturn's radius, the hydrogen is in this liquid metallic state. Jupiter also has a liquid metallic hydrogen mantle, but on that planet this layer is far larger than on Saturn. As with Jupiter, the magnetohydrodynamics of Saturn's liquid metallic hydrogen is the source of the planet's magnetic field. Planetary magnetic fields are believed to be produced through the motion of a highly conducting fluid through a magnetic field. The conductor then reinforces the existing field, creating a rather stable and permanent planetary field. This is the dynamo model of planetary magnetic fields. Because Saturn's liquid metallic hydrogen layer is far smaller than Jupiter's, Saturn's magnetic field is much weaker than Jupiter's, being only about 3 percent as strong. Even so, it is still nearly six hundred times stronger than Earth's magnetic field.

Saturn, like Jupiter, radiates more energy than it gets from the Sun, almost 2.9 times what it gets from the Sun. Jupiter radiates energy through kelvin-Helmholtz contraction, a form of hydrostatic contraction that permits a fluid body to compress, generating thermal energy. Saturn has much less mass than Jupiter. Theory indicates that kelvin-Helmholtz contraction is unable to produce the level of thermal energy needed to account for observations of Saturn. However, Saturn's temperature is lower than Jupiter's. At the temperature and pressure of Saturn's liquid hydrogen layers, helium, another major component of Saturn, precipitates out. Droplets of helium, heavier than hydrogen, sink toward the planet's deeper interior. The sinking helium releases gravitational energy in the form of heat, warming the planet. This thermal energy is nearly

## Saturn Compared with Earth

| Parameter | Saturn | Earth |
|---|---|---|
| Mass ($10^{24}$ kg) | 568.46 | 5.9742 |
| Volume ($10^{10}$ km³) | 82,713 | 108.321 |
| Equatorial radius (km) | 60,268 | 6,378.1 |
| Ellipticity (oblateness) | 0.09796 | 0.00335 |
| Mean density (kg/m³) | 687 | 5,515 |
| Surface gravity (m/s²) | 10.44 | 9.80 |
| Surface temperature (Celsius) | −160 | −88 to +48 |
| Satellites | 60 | 1 |
| Mean distance from Sun millions of km (miles) | 1,434 (884) | 150 (93) |
| Rotational period (hrs) | 10.656 | 23.93 |
| Orbital period (days) | 10,760 | 365.25 |

*Source:* National Space Science Data Center, NASA/Goddard Space Flight Center.

double the energy that the planet gets from the Sun, resulting in Saturn's observed emissions.

## Knowledge Gained

Most observations of Saturn have, of necessity, been made from Earth. The interior of the planet, however, cannot be studied from afar. Spacecraft sent to Saturn have yielded more data, but there is still considerable debate about the nature of Saturn's interior. Mathematical models of Saturn's interior are made based on observed characteristics, but more research is needed. The exact nature of Saturn's interior is therefore still the object of much speculation.

Because Saturn is shrouded in clouds, astronomers are unable to measure Saturn's rotational rate directly using observations from Earth. The planet's magnetic field should rotate with the planet. In 1981, Voyager 2 measured a rotational rate of a bit over 10 hours, 39 minutes. In 2007, however, scientists at the National Aeronautics and Space Administration's (NASA's) Jet Propulsion Laboratory released a new finding of 10 hours, 32 minutes, and 35 seconds for Saturn's rotational rate using data from the Cassini orbiter. The discrepancy between these measurements may be explained by the fact that it is unusually difficult to determine the rotational rate for Saturn because the planet's magnetic axis is nearly the same as its rotational axis, providing little change in the magnetic field for spacecraft to measure as the planet rotates.

Saturn and Jupiter both likely formed from the cloud of gas swirling together to form the Sun. Thus, these two gas giants would be expected to have very similar compositions to the Sun. Indeed, like the Sun, Jupiter and Saturn are both composed mostly of hydrogen and helium. Saturn's atmosphere, however, contains quite a bit less helium than either the Sun or Jupiter. This finding was a mystery until astronomers found that Saturn radiates far more energy than it gets from the Sun. The explanation that helium precipitating to lower levels of Saturn could heat the planet also explains why the upper portions of Saturn, which can be observed, are deficient in helium.

Hydrogen and helium are the primary constituents of Saturn. Both Jupiter and Saturn are somewhat enriched in elements heavier than hydrogen and helium when compared with the Sun. This is expected because both planets have, over the several billion years since the planets' formation, accreted planetesimals, asteroids, and comets. Saturn is considerably more enriched

in these heavier elements than is Jupiter, however, and the reason is unclear. Saturn may have simply accreted a greater percentage of these bodies, or it may have collected less hydrogen and helium when it formed. Further research is needed to answer the question of why Saturn and Jupiter have this compositional difference.

Studies of Saturn and Jupiter have led some astronomers to theorize that these planets may have formed somewhat closer together than they currently are in the solar system. Jupiter migrated closer to the Sun, and Saturn migrated somewhat farther from the Sun, to its current position.

## Context

The two largest planets in the solar system, Jupiter and Saturn, probably formed in the same manner, at about the same time, in the same part of the disk of material swirling together to form the Sun. Thus, they would be expected to be very similar. Indeed, they are similar, but there are important differences between them. Understanding those differences will cast light on the conditions under which gas giant planets form. In turn, understanding the formation of the gas giants will improve our understanding of the formation of other planets in the solar system, such as Earth, as well as the planetary systems of other stars.

The planet Saturn, because of its great distance from Earth—more than 1.2 billion kilometers at its closest—has mostly been studied via telescopes here on Earth. However, four spacecraft have investigated Saturn up close. The first was Pioneer 11, which passed closest to Saturn on September 1, 1979. Then, two Voyager spacecraft visited Saturn, with Voyager 1 passing closest to Saturn on November 12, 1980, and Voyager 2 flying past Saturn on August 25, 1981. No spacecraft visited Saturn until more than two decades later, when the Cassini orbiter entered orbit around Saturn on July 1, 2004. Cassinni has been studying Saturn and its satellites ever since. Although Cassini carried the Huygens probe to study Saturn's satellite Titan, no atmospheric probe was carried to study Saturn's atmosphere or interior, so all studies of Saturn's interior must be made by inference from observations of Saturn's exterior and of the planet's magnetic field. This means that scientists do not yet have a firm grasp of Saturn's interior, and further research is needed to understand this planet's interior structure fully.

*Raymond D. Benge, Jr.*

## Further Reading

Bortolotti, Dan. *Exploring Saturn*. New York: Firefly Books, 2003. A look at the Cassini-Huygens mission for a younger audience. Full of charts, photographs, a section on observing Saturn, and discussion of the history of our knowledge of the Saturnian system from antiquity to the launch of Cassini.

Corfield, Richard. *Lives of the Planets: A Natural History of the Solar System*. New York: Basic Books, 2007. A history of planetary exploration that emphasizes the exploration more than the planets themselves.

De Pater, Imke, and Jack J. Lissauer. *Planetary Sciences*. New York: Cambridge University Press, 2001. An upper-division or graduate-level textbook on planetary sciences, with a short description of the interior structures of gas giant planets. This text requires a fairly high level of mathematical familiarity.

Freedman, Roger A., and William J. Kaufmann III. *Universe*. 8th ed. New York: W. H. Freeman, 2008. An excellent college-level introductory astronomy textbook. An entire chapter is devoted to Jupiter and Saturn.

Harland, David M. *Cassini at Saturn: Huygens Results*. Chichester, England: Praxis, 2007. This book is mostly about Saturn's moons, but it also gives a good description of data from the Cassini orbiter.

Irwin, Patrick. *Giant Planets of the Solar System: An Introduction*. New York: Springer, 2006. An overview of all four gas giants in the solar system, this text is written at a level for advanced students. It covers all aspects of the planets, including theories of formation, and has a very good bibliography.

Lovett, Laura, Joan Harvath, and Jeff Cuzzi. *Saturn: A New View*. New York: Harry N. Abrams, 2006. A coffee-table book with about 150 of the best images returned by the Cassini mission to Saturn. Covers the planet, its many satellites, and the complex ring system.

Morrison, David. *Voyages to Saturn*. NASA SP-451. Washington, D.C.: Government Printing Office, 1982. A good overview of the Pioneer and Voyager missions to Saturn, and some of their initial findings, with numerous illustrations. Some of the information is dated, but still useful.

# SATURN'S MAGNETIC FIELD

**Categories:** Planets and Planetology; The Saturnian System

*The magnetic field of Saturn, which was discovered and first analyzed from data collected by Pioneer 11, Voyager 1, and Voyager 2, reveals similarities to the magnetic fields of both Earth and Jupiter yet has many distinctive features. The Saturnian magnetosphere provides yet another cosmic laboratory for the study of astrophysically important processes such as collision-free shocks, charge-accelerating processes, plasma-wave modes, particle trapping, diffusion, and a host of related phenomena. The Cassini spacecraft provided an orbital platform from which those processes were studied on a regular basis for many years beginning in 2004.*

## Overview

After it was established in the early 1960's that decimetric radio emissions of Jupiter, discovered in 1955, were emitted by high-energy electrons trapped in the giant planet's intense magnetic field, a search for the scaled-down version of similar radiation from Saturn led to observation of radio noise from Saturn. Subsequently, a first model of Saturn's magnetic field was developed. Cassini observations resulted in the greater fidelity of that model and revealed that Saturn's satellite Enceladus possesses its own magnetic field and emits radio waves of a complex nature.

Initial concrete evidence for the existence of the Saturnian magnetic field was provided by Pioneer 11's magnetometer. The bulk of Saturn's magnetic field is believed to be generated by rapid internal motion of the metallic hydrogen that surrounds the planet's rocky central core, forming a dynamo and resulting in a field that resembles that of Earth and Jupiter. Field strength at cloud-top level over Saturn's equator was found to be 0.2 gauss, roughly a third of the equatorial geomagnetic field. By comparison, Jupiter's magnetic field at cloud-top level is ten times stronger than that of Earth. Saturn's magnetic axis and rotational axis are nearly coincident. The field is believed to originate at a greater depth than do those of Earth and Jupiter in relation to their respective radii. Moreover, the rotational and magnetic axes of both Earth and Jupiter make a nearly 11° angle with their respective rotational axes. Pioneer 11 and Voyager data indicated that the center of Saturn's magnetic dipole axis is offset by 4 percent of Saturn's radius to the north of the center of the planet, and the polarity of the field is reversed with respect to Earth's polarity. These facts combine to give Saturn's external magnetic field a perfect symmetry, with none of the wobbles that characterize the fields surrounding Earth and Jupiter.

Another component of Saturn's magnetic field is an extensive ring of current of 10 million amperes flowing from west to east. Near the axis, the ring-current field is not parallel to the main field but becomes parallel with distance. Moreover, field lines are outward bound above the equatorial plane and inward bound below that plane. Thus, Saturn's magnetic field is made up of the main dipole field and the ring-current field, along with the boundary currents of the magnetopause (the border at which the solar wind meets the planetary magnetic field), thereby defining the magnetosphere. Because of the superposition of the ring-current field on the planet's intrinsic dipole field, the shape of the overall magnetosphere is stretched outward along the equatorial plane, creating a bulge that distorts the poloidal aspect of the field.

As confirmed by Pioneer 11, the Voyagers, and the Cassini orbiter, interaction between Saturn's magnetic field and the solar wind causes a magnetosphere accompanied by a well-defined bow shock, a plasma sheath,

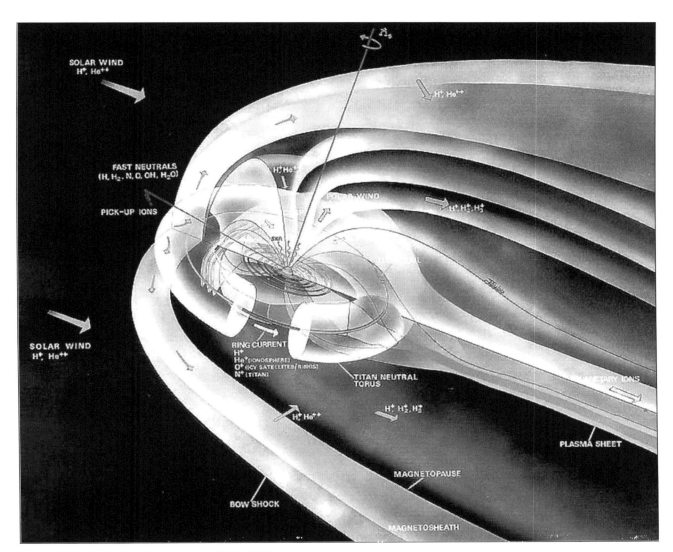

*An artist's diagram of Saturn's magnetic field.* (ESA)

and a magnetotail, similar in many respects to those associated with Earth and Jupiter. The shape of the magnetosphere can be schematically represented by a paraboloid of revolution around the Saturn-Sun line. The size of the magnetosphere depends on the varying pressure of the solar wind. If this pressure is low, the magnetopause occurs at a greater distance, thus inflating the magnetosphere. The solar activity cycle, interplanetary conditions, and local variations influence solar wind pressures. Magnetospheric and myriad complex radiative processes that occur within it result primarily from the interaction between charged particles of the solar wind and Saturn's intrinsic magnetic field. It is assumed that the transfer of energy from the solar wind to the magnetosphere occurs within the outer layer of the magnetosphere and the magnetotail. The magnetosphere effectively entraps, stores, and reradiates the energy of the solar wind.

With regard to Saturn's magnetic field, Pioneer 11 and the Voyager probes found that there was a near alignment of the magnetic and rotational axes and a displacement of the magnetic dipole center compared to the planetary center. They also discovered the existence of an additional field produced by a ring of current flowing west to east in the equatorial plane. Because of the varied areas covered by these probes' differing trajectories and the complementary data they provided, the Pioneer and Voyager missions provided a fairly accurate model of the Saturnian magnetic field configuration. Cassini's magnetometer filled in a great many gaps and found some surprises, especially with regard to the satellite Enceladus.

Saturn's ring systems are intricate, elaborate, and fascinating both in appearance and from a scientific point of view. The rings are thin and known to be composed of pieces of ice and chunks of rocky material, most of which are probably much smaller than ten meters in diameter. Stars can dimly be seen through the rings. The entire system rotates in the equatorial plane, within about 180,000 kilometers of the center of the planet. All of Saturn's major satellites, starting with Mimas, are located beyond the principal ring systems. Smaller ones are embedded and inside the ring systems.

The regions surrounding the major rings of Saturn are devoid of trapped charged particles. Up to a radial distance of about 150,000 kilometers, the density of charged particles is found to be negligible. Thereafter it rises abruptly, with intermittent variations caused by the absorption of particles by the satellites and the tenuous E ring. The maximum density of ions and electrons is observed between the orbits of Tethys and Rhea, between

300,000 and 60,000 kilometers from Saturn's center. The same region, which is rich in dense plasma, is lacking in low-energy electrons and protons. They are most likely absorbed by the neighboring satellites.

In this inner magnetosphere, spinning rapidly, charged particles are coupled to the magnetic field, resulting in a 60,330-kilometer-thick plasma sheet about 240,000 to 422,000 kilometers from the planet. From this distance to the orbit of Titan, there is a vast torus of neutral hydrogen, produced by the photochemical breakdown of methane escaping from the giant satellite's atmosphere. Another source of charged particles is provided by the collision of neutral hydrogen atoms with the magnetosphere of Titan.

Saturn's magnetospheric radio emission has three components, all of which are presumed to originate in the magnetic plasma surrounding the planet. Its kilometric radiation was found to have a fundamental periodicity of 10 hours and 40 minutes, presumably the rotational period of the magnetosphere. These magnetic storms seem to affect and are correlated with solar wind conditions, as well as the relative position of the satellite Dione. Saturn's kilometric radio emissions, first discovered by Voyager 1 in January, 1980, exhibit a wide range of period modulation (10.66 hours, 66 hours, and 22 days) along with intensity and power variations. The axial symmetry of Saturn's magnetic field does not yield the kind of intensity modulation that is associated with the wobbly motion of Jupiter's field. Saturn appears to radiate greater power toward its dayside than toward the nightside; this phenomenon has its origin in midlatitude to polar cusps, regions where magnetic field lines enter or leave the planet. Magnetospheric kilometric radiations—short bursts of radio emissions caused by electrostatic discharges ranging in frequency from 20 kilohertz (kHz) to 40 megahertz (MHz) with a periodicity of 10 hours and 10 minutes—were also detected by the planetary probes. The total power radiated by these discharges appears to be comparable to those of the kilometric emissions. The plasma trapped within the magnetosphere also has radio emissions, with a frequency exceeding 2 to 3 kHz and a recurrence period of 10 hours and 40 minutes. These low-frequency radio waves were detected by both Voyager probes. The total power radiated is estimated to be on the order of a million watts.

A dense atmosphere combined with a strong magnetic field indicates the presence of auroras. Voyager 1's detection of strong ultraviolet radiation above 76° north latitude and below 78° south latitude confirms the existence of Saturnian auroras similar to those in the polar regions of Earth and Jupiter, with possible periodic longitudinal variation in intensity. Based on these measurements, the power required to generate the Saturnian auroras is estimated at 200 billion watts, which is seven times the corresponding value for Earth's auroras. The main source of this energy is the interaction between the solar wind and Saturn's magnetosphere. Confinement of charged particles within the magnetic field and their being forced to travel along intense magnetic field lines near the polar regions cause the auroras. More intense study of Saturn's auroral activity was effected using the Hubble Space Telescope and from the orbiting Cassini spacecraft.

Because Saturn's magnetic field is highly symmetrical, there is a well-defined bow shock and a steady magnetopause. Saturn has the necessary ingredients for the presence of an active outer magnetic field: a planetary dipole field, a ring-current field, and contributions to the field from a magnetopause and tail currents. The outer magnetosphere of Saturn is supplied with a constant flow of hydrogen and nitrogen plasma by Titan and its magnetic field. This plasma torus is co-rotating with the magnetosphere, undergoing convective motion at the same time, and is estimated to have a temperature of 1 million kelvins—the hottest of planetary surroundings known to date. Neutral atoms escaping from Titan are photoionized, becoming part of the plasma torus. It is surmised that the co-rotations of plasma, combined with the continual radial movement of plasma torus caused by fluctuations in the pressure of the solar wind, are responsible for the plasma's unusual heating, which leads to an increase in high-energy particles. In addition to the elevated plasma temperature that results in high-energy particles, there is a coincident rise in field strength. This jump in field strength appears to be sharper when the magnetosphere contracts. Voyager data indicate discontinuities between the region inside the magnetosphere and the outside, where the solar wind persists. An acceleration of electrons and ions occurs in the magnetotail. Voyager 1 detected low-energy ions streaming toward, and high-energy ions racing away from, Saturn at distances of 2 million to 2.7 million kilometers.

The bow shock, the magnetopause, the acceleration of ions in the magnetotail, a spectrum of energy, and the flux of charged particles appear to be common to the outer magnetospheres of Saturn and Earth. A major difference is the Titan-fed hot plasma torus of Saturn's outer magnetosphere. There are also many unexplained phenomena in Saturn's magnetosphere, one of them being the observed fluctuation in temperature at altitudes between 600,000 and 900,000 kilometers. It is not known whether the magnetotail drains the angular momentum of Saturn through ejection of high-energy particles.

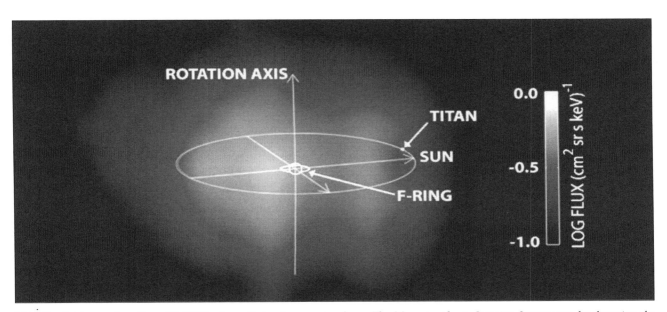

*This Cassini image from June 21, 2004, shows Saturn's magnetosphere: The Magnetospheric Imaging Instrument, by detecting the hydrogen atoms that escape the magnetosphere, revealed a glow that reflects the area of magnetospheric influence around the planet.* (NASA/JPL/Johns Hopkins University)

## Knowledge Gained

The search for Saturn's magnetic field began in 1955, with the accidental discovery of decimetric radio emissions from Jupiter. It was surmised that radio emissions from Saturn's magnetic field would be weaker than those of Jupiter and hence not observable by ground-based radio telescopes. Long-wave radio bursts coming from the direction of Saturn were detected in 1975 by the Interplanetary Monitoring Platform (IMP) 6. Instruments on board Pioneer 11 in 1979, however, confirmed the existence of an extensive magnetic field surrounding Saturn. Detailed observations were carried out by more sophisticated instrumentations of Voyagers 1 and 2. Long-term studies of the magnetic fields and particle environment around Saturn continue to be performed by the Cassini spacecraft.

Data from Pioneer 11 and the two Voyagers suggest that the magnetic field of Saturn originates at a greater depth than does Jupiter's. Unlike the fields of Earth and Jupiter, Saturn's magnetic field is highly axisymmetric because of the overlapping of the dipole and the rotational axes. The main field is hemispherically asymmetrical because of a northward offset of the dipole axis. Field strength at an altitude of 60,330 kilometers is 0.2 gauss, roughly a third of the equatorial geomagnetic field. A second component of the magnetic field of Saturn is the extensive current ring that flows from west to east. The boundary currents of the magnetopause constitute a third component of the field.

The inner magnetosphere of Saturn has been found to be free of charged particles. The maximum density of ions and electrons occurs between the orbits of Tethys and Rhea. Between the orbits of Tethys and Titan, there is a vast torus of neutral hydrogen produced by the photochemical breakdown of methane escaping from Titan's atmosphere. Saturn's kilometric radio emission, along with a variety of periodic and nonperiodic bursts of radiation, is presumed to originate in the magnetic plasma surrounding the planet. The planet's kilometric radiation has a periodicity of 10 hours and 40 minutes. Short bursts of radio emissions have been traced to electrostatic discharges.

Pioneer 11 and the two Voyagers also provided tantalizing information about Saturn's auroras, phenomena that occur invariably in the presence of a dense atmosphere combined with a strong magnetic field. Saturn's auroras, which occur above 76° north latitude and below 78° south latitude, are powered by the interaction between the solar wind and the magnetosphere.

Cassini's magnetometer naturally was capable of determining the direction and strength of Saturn's magnetic field with the best technology available at the time the spacecraft was designed. Called a dual technique magnetometer, it also was able to assist in the determination of the size and nature of Saturn's core. In order to have the sensitivity necessary for its intended research, the magnetometer assembly included a flux gate magnetometer and a vector/scalar helium magnetometer placed along an 11-meter-long boom to avoid interference from spacecraft electronics. The ability to determine three-dimensional magnetic field maps would be used not only on Saturn but also to search for magnetic fields of the ringed planet's satellites. The magnetometer was intended to make new discoveries and to answer some of the outstanding questions raised by the Voyager results.

Cassini confirmed the Pioneer 11 and Voyager magnetic field findings, also returning some surprising insights about Saturn's satellite Enceladus. Much remains to be learned about this complex miniature-solar system during Cassini's extended mission, which was authorized in 2008, given the orbiter's good health, after Cassini's primary mission ended.

## Context

Study of Saturn's magnetic field began with radio astronomers, whose findings and conjectures were to play a dominant part in the planning of planetary exploration probes of the 1970's. Pioneer and Voyager data vastly advanced scientists' understanding of planetary magnetic fields in general and Saturn's field in particular. For example, the axisymmetric nature of Saturn's magnetic field and its effects on magnetospheric processes make it quite different from both Earth's and Jupiter's fields. However, data show that Saturn's magnetosphere (like those of Earth, Jupiter, and Venus) has an extended magnetotail, in which interaction between solar plasma and the magnetic field produces a host of phenomena yet to be researched. The study of magnetospheric processes in general provides a basis for understanding cosmic plasma.

Saturn's magnetic field is basically dynamo-driven, like similar fields in the solar system, including even that of the Sun. Close to the planet, plasma depletion is caused by the ring system closer to the planet, but there is a vast torus of Titan-fed hot plasma at the outer magnetosphere. Saturn appears to possess three distinct regions of plasma: an inner plasma torus, an extended plasma sheet, and the hot outer plasma torus. Both the temperature and the thickness of the plasma disk increase with distance from the planet. The nature of the sources and sinks of these plasma regions is yet to be determined, although speculations abound. The plasma torus is co-rotating, with velocities decreasing by 10 to 20 percent beyond 480,000 kilometers of altitude.

Based on knowledge of Earth's magnetic field and the data provided by Pioneer 11, Voyagers 1 and 2, and Cassini, a model of Saturn's field has been constructed. There are undoubtedly numerous processes occurring in its distant magnetosphere that need to be further examined. The variation in the size and shape of the magnetosphere needs to be examined over an extended period. Quantitative studies of the plasma flow from the satellites of Saturn (Titan and Enceladus in particular) and research into the charge absorption properties of the ring system are projects of long-term interest as well.

The theoretical model of Saturn's magnetic field is based on a solid foundation provided by spacecraft data. Yet there are numerous unanswered questions and doubts, and the model must be further refined. Continuing research into the atmospheres and magnetic fields of planets such as Jupiter and Saturn is vital. However, no follow-on mission to Saturn beyond Cassini was advanced and approved by the time that Cassini's flight was extended two additional years after completion in 2008 of the orbiter's primary mission objectives.

*V. L. Madhyastha*

## Further Reading

Alexander, Arthur Francis O'Donel. *The Planet Saturn: A History of Observation, Theory, and Discovery.* New York: Macmillan, 1962. An excellent historical reference for all readers, this book details the study of Saturn up to the mid-twentieth century. Organization is somewhat choppy, however, rendering some sections difficult to read.

Bortolotti, Dan. *Exploring Saturn.* New York: Firefly Books, 2003. A look at the Cassini-Huygens mission for a younger audience. Full of charts, photographs, a section on observing Saturn, and a history of our understanding of the Saturn system from antiquity to the launch of Cassini.

Gehrels, Tom, and Mildred Shapley Matthews, eds. *Saturn.* Tucson: University of Arizona Press, 1984. Many of the articles collected here are by individuals who were involved with the Pioneer 11 or Voyager 1 and 2 projects; some of them were also responsible for publishing the first reports of the projects' scientific findings. Thus, the work is authoritative, and though it

is intended for the specialist, the general reader can obtain from it much useful information, including helpful lists of references.

Harland, David M. *Cassini at Saturn: Huygens Results.* New York: Springer, 2007. The cover illustrates a landing site on Titan. The text inside provides a thorough explanation of the entire Cassini program, including the Huygens landing on Saturn's largest satellite. Essentially a complete collection of National Aeronautics and Space Administration's releases from the start of Cassini flight operations through the majority of Cassini's seventy orbits during its primary mission, which concluded a year after this book was published. Technical but accessible to a wide audience.

_____. *Mission to Saturn: Cassini and the Huygens Probe.* New York: Springer Praxis, 2002. A technical description of the Cassini program, its science goals, and the instruments used to accomplish those goals. Written before Cassini arrived at Saturn, it nevertheless provides a historical review of pre-Cassini knowledge of the Saturn system.

Hartmann, William K. *Moons and Planets.* 5th ed. Belmont, Calif.: Thomson Brooks/Cole, 2005. An updated version of a classic text that covers all aspects of planetary science. Covers the Pioneer 11 and Voyager data concerning Saturn and previews Cassini mission objectives. Takes a comparative planetology approach rather than approaching each moon and planet in separate chapters.

Irwin, Patrick G. J. *Giant Planets of Our Solar System: An Introduction.* 2d ed. New York: Springer, 2006. Suitable as a textbook for upper-level college courses in planetary science, this volume focuses on Jupiter, Saturn, Uranus, and Neptune and their satellites, rings, and magnetic fields. Filled with figures and photographs, the work is accessible to the serious general audience.

Morrison, David. *Voyages to Saturn.* NASA SP-451. Washington, D.C.: Government Printing Office, 1982. This is a well-written account of the encounters of Voyagers 1 and 2 with Saturn and a summary of the resulting scientific findings. Containing a wealth of information that is probably not available to the general reader elsewhere, this volume is a primary resource for Voyager returns on Saturn.

Russell, Christopher T. *The Cassini-Huygens Mission: Orbiter Remote Sensing Investigations.* New York: Springer, 2006. Provides a thorough explanation of the remote-sensing investigations of the orbiter and lander.

Outlines the scientific objectives of all instruments on the spacecraft and describes the planned forty-four encounters with Titan. Given the publication date, only early science returns are covered.

Schardt, A. W. "Magnetosphere of Saturn." *Review of Geophysics and Space Physics* 21 (1983): 390-402. This historic review article briefly describes Saturn's magnetic field and its associated phenomena in the light of findings from Pioneer 11, Voyager 1, and Voyager 2. The author presents a brief history of the projects and assumes minimal technical knowledge on the part of the reader. The article includes an extensive list of references.

# SATURN'S RING SYSTEM

**Categories:** Planets and Planetology; The Saturnian System

*Data transmitted by Pioneer 11, and in greater detail by Voyagers 1 and 2, revolutionized the understanding of Saturn's complex ring system previously obtained by observations from Earth-based telescopes This information revised models based on those earthbound investigations dating back to the discovery of the rings more than three centuries ago. Hubble Space Telescope and Cassini orbiter studies then built upon data returned by earlier spacecraft and revealed an even more complex ring system at Saturn.*

## Overview

Pioneer 11, Voyager 1, and Voyager 2—three deep space probes launched between 1973 and 1977—all encountered Jupiter before their trajectories were directed toward Saturn. Pioneer 11 was also known as Pioneer Saturn during its Saturn flyby, as it was a pathfinder for the more sophisticated two Voyager spacecraft. Voyager 2, launched after Pioneer, would have to pass through Saturn's ring plane at a distance of 2.86 Saturn radii (about 112,000 kilometers above the surface) in order to be put on a trajectory for a potential Uranus flyby. Although beyond the main rings, there was the possibility that a tenuous ring existed in this region. Such a ring would pose a threat to any spacecraft passing through at high speed. The decision was made to have Pioneer 11 cross the ring plane at this distance to determine whether it was safe for the more valuable Voyager spacecraft to come.

The Cassini spacecraft was launched on October 15, 1997, and after a series of gravity assists arrived in orbit about Saturn on July 1, 2004. The primary mission of this spacecraft was completed in four years, but fortunately funding was available for an extended examination of the Saturn system as the spacecraft remained in near-perfect health at the end of its primary mission.

Pioneer 11 survived the crossing of the ring plane on September 1, 1979, at a distance of 2.82 Saturn radii. The success of this maneuver was evident from the continued reception of the spacecraft's radio transmissions before, during, and after the ring crossing. Pioneer 11 then swung around the planet, crossing the ring plane a second time (about 2.5 hours after the first crossing) at a distance of 2.78 Saturn radii. Again, although it took some hits, there was no detectable damage during this crossing.

During its Saturn flyby, Pioneer 11 transmitted images of the rings made by an imaging photopolarimeter, an instrument which produces images by means of polarized light. In this case it made images at two visible wavelengths, one in the red region of the electromagnetic spectrum and another in the blue. These images could be processed at the receiving station to simulate color pictures. As the spacecraft approached Saturn, the resolution exceeded that of earthbound observations. At a distance of about 1 million kilometers, a new ring was detected in one of the images. It was named the F ring. It was too narrow and too close to the outer edge of the A ring to have been seen from Earth. Pioneer 11 also collected data on Saturn's rings by using transmitted sunlight, a method not possible for observations made from Earth. With this method, images are analogous to a photographic negative; gaps in the rings appear bright instead of dark, while the dense, bright parts of the rings appear dark. A week after the first encounter, the Saturn flyby was essentially complete, and Pioneer 11 was on its way on an escape trajectory that would take it out of the solar system.

As the planet moves about the Sun, the orientation of the ring plane relative to Earth varies. In early 1980, prior to the Voyager 1 flyby, Saturn's rings appeared sideways in Earth-based observations. The main rings were practically invisible, because they appeared so thin from that perspective. Conditions were favorable, however, for detecting faint satellites and diffuse rings which would otherwise be lost in the glare of the main rings. A faint ring was detected beyond the F ring by astronomers during this period. This ring, now called the E ring, had actually been discovered in 1966, the last time that the rings were edgewise, by a then-unknown astronomer, Walter A.

*Saturn and its rings, compiled from thirty-six images taken as the Cassini spacecraft approached on January 19, 2007, about 1.23 million kilometers from the planet. The planet casts a shadow on the rings facing away from the Sun.* (NASA/JPL/Space Science Institute)

Feibelman. At that time, its existence was called into question by other observers. Pioneer 11 vindicated Feibelman.

The Voyager imaging system represented a tremendous improvement over that of Pioneer 11. It consisted of two television cameras. One was outfitted with a wide-angle lens while the other had a narrow-angle lens. These cameras were mounted on a scan platform which could be aimed continuously at the target, a process referred to as slewing or motion compensation. This prevented smearing of images due to the rapid motion of the Voyager spacecraft, and was accomplished without using propellant to reorient thrusters to move the spacecraft. Pioneer 11's imaging photopolarimeter had to rotate with the spin-stabilized spacecraft, and could only record the subject once during each revolution. Voyager images could be transmitted

more rapidly in black and white or in color. They also had higher resolution.

During October, 1980, Voyager 1 was traveling toward Saturn at an average rate of 1.3 million kilometers per day. At the end of the month it was 17 million kilometers from the planet, and the improved resolution brought to light new details within the rings. The principal rings, particularly the B ring and C ring, were found to consist of narrow concentric rings. The ringlets, as they were called, proved to be so numerous that they suggested the analogy of grooves on a phonograph record. The main gap in the ring system, the Cassini division between the A ring and B ring, was not a total void. It contained some ringlets of its own.

The B ring exhibited some curious radial streaks referred to as spokes which rotated with the ring while changing their shapes. This was a completely baffling phenomenon. In order to study them more carefully, Voyager 1 was programmed to make images every five minutes over a period of ten hours. In the process, two small satellites were discovered. One orbited along the outer edge of the F ring, the ring that had been discovered by Pioneer 11, while the other orbited along the ring's inner edge. The specific location of these satellites is considered to be essential to the permanence of the F ring. The F ring itself also has a curious property. It is braided.

During its encounter with Saturn's rings on November 12, 1980, Voyager 1 swung around Saturn and crossed the ring plane twice without damage. Before being directed to escape the solar system, it transmitted spectacular images of Saturn and its rings. The number of ringlets that could be detected with the Voyager instrumentation was estimated to have an upper limit of one thousand. Voyager 1 discovered two more rings, the D ring, which replaced the C ring as the innermost ring, and the G ring, beyond the F ring discovered by Pioneer 11. Evidence for the G ring had been obtained by Pioneer 11. Its charged particle

detector registered a decrease in intensity which could have been caused by absorption by the ring particles. Voyager 1 confirmed the existence of the E ring by photographing it directly.

As a result of discoveries made by Voyager 1, the program of Voyager 2 was revised for focused studies of the rings. A decision was made to program the second Voyager to collect data during a stellar occultation of Saturn's rings. During the occultation, Voyager 2 would be positioned above the ring plane, focusing on the selected star through Saturn's rings. Delta Scorpii, the star, was bright enough to be seen through the partially transparent rings. One of Voyager's instruments, the photopolarimeter subsystem (PPS), would be focused on the star for about two hours. The PPS would not form images. It would measure the rapid fluctuations in the brightness of the star. These fluctuations would be caused by the ring particles in the line of sight momentarily blocking the light from the star. Data could be analyzed to map variations in ring

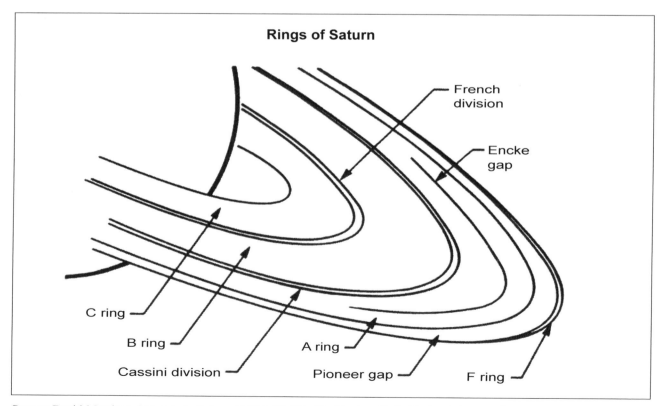

**Rings of Saturn**

French division

Encke gap

C ring

B ring

Cassini division

A ring

Pioneer gap

F ring

*Source:* David Morrison, Voyages to Saturn. NASA SP-451. Washington, D.C.: National Aeronautics and Space Administration, 1982, p. 25.

structure. Voyager 2 also scanned the area for small satellites which might be embedded within the rings.

As Voyager 2 approached Saturn, images of the rings achieved an extraordinary resolution of ten kilometers. Still, there was no evidence of embedded satellites. The number of observed ringlets increased with the improved resolution. On August 25, 1981, the occultation of Delta Scorpii was successfully observed. As seen from Earth, Voyager then proceeded to cross the ring plane behind Saturn. The successful crossing could not be confirmed until radio communications were resumed about one hour later. Voyager 2 was then directed toward Uranus. Shortly afterward, ground controllers discovered that the scan platform had lost its azimuth motion and had been shut down by the on-board computer. Many of the programmed images had not been acquired, although some good images of the F ring were obtained before the scan platform failed. Quick actions by the Voyager team salvaged much of the anticipated scientific investigations as Voyager 2 left Saturn.

As Cassini approached orbit about Saturn, it made its closest planned encounter with the planet's rings. Looking at the B ring Cassini noted a puzzling lack of spokes presently. This strongly suggested that the spoke phenomenon might be a seasonal effect. Later in the mission Cassini produced numerous images of the spokes.

Cassini also examined gaps in the rings and noted far more structure in them than had the Voyagers, and found additional shepherding satellites. For example, inside the 42-kilometer-wide Keeler gap within the A ring, in May, 2005, Cassini discovered a small satellite. This body clears out material within this gap. In 2006 Cassini imaged a very faint dust ring located near the satellites Janus and Epimetheus. It is believed that this 5000-kilometer-wide ring is composed of particles liberated by meteoroids impacting these two satellites. That same year another faint dust ring was found, this one existing close to the satellite Pallene. This one is only 2500 kilometers in radial extent. It too is suspected to be composed of particles generated by collisions of objects, specifically with Pallene.

**Knowledge Gained**

Three of Saturn's major rings (D ring, F ring, and G ring) were discovered with the help of deep space probes. The existence of the faint E ring was confirmed. The rings, in order of increasing distance from the planet, are D, C, B, A, F, G, and E. Saturn's rings were first named in alphabetical order in 1850: A, B, and C for the three rings then

known. The convention now is to name the rings in the order of their discovery.

Three narrow gaps in the rings were also found: one in the C ring, one at the inner edge of the Cassini division, which separates the A ring and B ring, and a third one in the A ring, close to its outer edge.

While individual ring particles have not been observed directly, the distribution of particle sizes can be estimated from measurements of the transmission of light and radio waves through the rings and scattering, or reflection, from the rings. Being an orbital platform, Cassini was able to conduct multiple radio occultation measurements. The sizes range from microns to about ten meters. Spectroscopic measurements confirm that these particles are primarily composed of ice, with some impurities. Considering the larger-sized particles only, scientists estimate that the thickness of the rings to be about one hundred meters. This is relatively thin since the ring system is thousands of kilometers wide.

Rings discovered by deep space probes differ from the ones first identified by ground-based observation. The outer two rings, G and E, are composed of particles in the micron range and do not have the ringlet structure. They are diffuse and are probably thicker than one hundred meters. It is believed that Voyager 2 passed through the G ring when it was leaving Saturn. Its plasma wave detector recorded a large increase in intensity in the vicinity of the ring. The increase was apparently caused by the spacecraft colliding with ring particles as it traveled through the ring at 10 kilometers per second. Impacts would have vaporized the particles, producing puffs of ionized gas (plasma) which would have been recorded by the detector.

Braided strands of the F ring observed by Voyager 1 were found by Voyager 2 to have changed to parallel strands. They changed back to braided strands by the end of the encounter, one example of the dynamic nature of the rings. Two so-called shepherd satellites serve to confine the ring. They are elongated rather than spherical. The long axis of the larger one is about 140 kilometers. Cassini was able to observe the dynamic behavior of the F ring over the course of its many orbits about Saturn.

Of all the ring components seen in the outer solar system, Saturn's F ring displays the most unusual and dynamic activity. Features in it can be seen to change on timescales ranging from just hours to several years. Studies using Cassini's Ultraviolet Imaging Spectrometer during stellar occultations, found at least a dozen objects within the F ring ranging from 27 to 10,000 meters in size. Data suggested that these features were aggregates that had temporarily clumped together, the supposition being that within the rings material coalesces and breaks apart due to gravitational and collision processes, respectively. In mid-2008 researchers using Cassini observations, published a paper in *Nature* that attributed the unusual characteristics of F ring dynamics to perturbations created by small moonlets, making the F ring the ring location in the solar system found thus far where a serious number of collisions happen regularly. In some ways F ring dynamics provide a window into the early solar system's protoplanetary disk at a time when collisions of small particles were needed to drive planetary formation.

In addition to the gravitational influence of the satellite Prometheus, a moonlet on the order of only a few kilometers in radius appears to be colliding with F ring particles orbiting close to the core of the ring generating fan-shaped structures. Another combination of gravitational effects and collisions produces structures referred to as jets. A Cassini image taken in 2004 appears to have identified a five-kilometer-wide moonlet that may produce some of the largest of the observed jets in the F ring. These

phenomena would continue to be studied throughout the extended Cassini mission.

The D ring appeared to be a collection of ringlets too faint to be seen from Earth. It has a relatively small percentage of micron-sized particles. Analysis of Voyager data revealed three ringlets. These were designated D68, D72, and D73 in increasing distance radially from the planet toward the C ring. Here again, Cassini revealed dynamic behavior and more complex structure. Twenty five years after its discovery, D72 was observed to be fainter and at a location 200 kilometers closer to Saturn. The gap between D73 and the C ring was not empty. Cassini recorded fine structure with ripples of material 30 kilometer apart.

The Cassini division, when viewed in transmitted light, appeared to have five rings in its central region and a gap on each side separating it from the A ring and B ring. The gap at the inner edge has an eccentric ringlet, which was scanned for small satellites that might confine it. None was found, but Cassini uncovered even more structure in the gap than revealed by the Voyagers.

## Rings of Saturn

| | Radius (km) | Radius/Eq. Radius | Optical Depth | Albedo ($\times 10^{-3}$) | Surface Density (g/cm$^2$) | Eccentricity |
|---|---|---|---|---|---|---|
| Saturn equator | 60,268 | 1.000 | — | — | — | — |
| D ring | >66,900 | >1.11 | — | — | — | — |
| C inner edge | 74,658 | 1.239 | 0.05-0.35 | 0.12-0.30 | 0.4-5 | — |
| Titan ringlet | 77,871 | 1.292 | — | — | 17 | 0.00026 |
| Maxwell ringlet | 87,491 | 1.452 | — | — | 17 | 0.00034 |
| B inner edge | 91,975 | 1.526 | 0.4-2.5 | 0.4-0.6 | 20-100 | — |
| B outer edge | 117,507 | 1.950 | — | — | — | — |
| Cassini division | — | — | 0.05-0.15 | 0.2-0.4 | 5-20 | — |
| A inner edge | 122,340 | 2.030 | 0.4-1.0 | 0.4-0.6 | 30-40 | — |
| Encke gap | 133,589 | 2.216 | — | — | — | — |
| A outer edge | 136,775 | 2.269 | — | — | — | — |
| F ring center | 140,374 | 2.329 | 0.1 | 0.6 | 0.0026 | — |
| G ring center | 170,000 | 2.82 | $1.0 \times 10^{-6}$ | — | — | — |
| E inner edge | ~180,000 | 3 | $1.5 \times 10^{-5}$ | — | — | — |
| E outer edge | ~480,000 | 8 | — | — | — | — |

*Source:* Data are from the National Aeronautics and Space Administration/Goddard Space Flight Center, National Space Science Data Center.

*This mosaic of six images from Cassini (taken December 12, 2004) displays gaps, gravitational resonances, and wave patterns across about 62,000 kilometers of the ring plane.* (NASA/JPL/Space Science Institute)

The B ring, the largest and brightest of the "classical" rings, has the most elaborate structure, literally thousands of ringlets. Spokes are transient features that appear bright in transmitted light but dark when viewed from the sunlit side. They consist of micron-sized particles. When electrically charged, they interact with Saturn's magnetic field, a process that explains some of their properties. There is evidence that the ringlets themselves are manifestations of a wave, propagating through the ring in the form of a spiral. This spiral might, in turn, produce the observed density variations. Cassini provided a great deal of evidence for not only density waves in the rings but also torsion waves.

## Context

History records that in 1610 Galileo was the first person to observe Saturn through a telescope. He described it as having a close satellite on either side. Later observers used the Latin word "ansae" (handles, in the sense of cup handles) to describe what they saw. This term is still used to refer to the parts of the rings that are visible on either side of the planet. Christiaan Huygens, based on observations in 1655, concluded that Saturn was surrounded by a thin ring not attached to the planet. Gian Domenico Cassini, discovered the main gap in the ring system in 1675, showing that there were two rings, now called A and B. The C ring was discovered in 1850.

In 1867, Daniel Kirkwood, an American astronomer, applied a resonance theory he had developed to explain the existence of the Cassini division. A ring particle in the division orbits Saturn with a period one-half that of Saturn's satellite Mimas, which is farther from the planet. Every other period, the particle passes Mimas in the same part of its orbit and is affected by a gravitational force pulling it outward. This periodic or resonant force would clear the ring of particles. The process is actually more complicated. Additional satellites have to be considered,

and the gravitational interaction is not as simple as described. In some instances, the force might produce a ringlet by causing a spiral density wave to form, as mentioned with respect to the B ring. Originally, this type of wave was introduced to explain the structure of spiral galaxies, such as the Milky Way. Just as the complexity of the ring system was completely unexpected, the concept that Saturn's rings may have some similarities to spiral galaxies, despite their enormous difference in size, could not have been anticipated before the Saturn encounters.

Prior to Pioneer 11, Voyagers 1 and 2, and Cassini, Saturn's rings could be described only in relatively simple terms. It had been established that the rings consist of discrete particles in orbit around the planet. Spectroscopic measurements showed that the inner parts of the rings rotate faster than the outer parts. Thus, it was clear that the rings were not solid disks.

In addition, it had been proved theoretically that any natural satellite large enough to be held together simply by the force of its own gravity would be fragmented by tidal forces exerted on it by the planet if it was closer than about 2.4 times the radius of the planet. This inner limit is known as the Roche limit, named for the nineteenth century French mathematician Edward Roche. All rings, except the F ring and E ring, are at a distance greater than 2.4 Saturn radii away from the planet. Roche suggested that the rings were formed by fragmentation of a satellite which came too close to the planet. Spacecraft data support another possibility. A number of Saturn's icy satellites were found to be pockmarked by impact craters. Ring particles could be remnants of the debris resulting from the collisions that produced such craters.

Voyager 2 successfully completed its Uranus encounter on January 24, 1986, and continued on its route to Neptune passing through that system in August, 1989. This spacecraft is unique in having made observations at close range of the four known planetary ring systems:

those of Jupiter, Saturn, Uranus, and Neptune. The extensive data set from the Saturn encounters made by four spacecraft form the basis for a unified model of planetary rings in general.

With the extension of the Cassini mission in 2008, there came the promise of new and exciting discoveries about the complex ring system of Saturn for several more years to come. One such unexpected investigation would be follow-up studies of a possible ring system about Saturn's large satellite Rhea. Cassini in November, 2005, was directed to use its Magnetospheric Imaging Instrument (MIMI) to observe the planet's magnetosphere in the vicinity of Rhea. MIMI found three specific diminishments of energetic particles located symmetrically about either side of this satellite, Saturn's second largest. The supposition was that there might be three rings made of particles on the order of a meter in size existing with an equatorial dish of material. This was not the only possible explanation, but if Cassini data confirmed it to be correct, then Rhea would be the only satellite known to have its own ring system.

*Howard L. Poss*

## Further Reading

Alexander, A. F. O. *The Planet Saturn: A History of Observation, Theory, and Discovery.* New York: Dover Publications, 1980. The standard work on the history of observations of Saturn, from ancient times to 1960. A large part of the book is devoted to the rings. The author frequently uses quotations from the original sources. A useful reference. Illustrated with drawings and photographs. Contains an extensive index.

Consolmagno, Guy. *Worlds Apart: A Textbook in Planetary Sciences.* Englewood Cliffs, N.J.: Prentice Hall, 1994. A text accessible to college-level science and nonscience readers alike. Presents subjects at low-level mathematics and also involves integral calculus where required. Demonstrates how the area of planetary science progresses by questioning previous understanding in light of new observations.

Elliot, James, and Richard Kerr. *Rings: Discoveries from Galileo to Voyager.* Cambridge, Mass.: MIT Press, 1984. The discovery of Uranus's rings is described. Other topics include the discovery of Jupiter's ring and the Saturn encounters. For college and high school students with background in the physical sciences. Includes an extensive bibliography of journal articles. Illustrated.

Encrenaz, Thérèse, et al. *The Solar System.* New York: Springer, 2004. A thorough exploration of the solar system from early telescopic observations through the space missions that have investigated all planets with the exception of Pluto by the publication date. Takes an astrophysical approach to give our solar system a wider context as just one member of similar systems throughout the universe.

Harland, David M. *Cassini at Saturn: Huygens Results.* New York: Springer, 2007. The cover illustrates a landing site on Titan. The text inside provides a thorough explanation of the entire Cassini program including the Huygens landing on Saturn's largest satellite. Essentially a complete collection of National Aeronautics and Space Administration (NASA) releases from the start of Cassini flight operations through the majority of Cassini's seventy orbits of its primary mission. Cassini's primary mission concluded a year after this book entered print. Technical writing style but accessible to a wide audience.

_____. *Mission to Saturn: Cassini and the Huygens Probe.* New York: Springer Praxis, 2002. Another book in Springer's Space Exploration Series, this is a technical description of the Cassini program, its science goals and the instruments used to accomplish those goals. Written before Cassini arrived at Saturn. Provides a historical review of pre-Cassini knowledge of the Saturn system.

Irwin, Patrick G. J. *Giant Planets of Our Solar System: An Introduction.* 2d ed. New York: Springer, 2006. Suitable as a textbook for upper-level college courses in planetary science. Focuses on Jupiter, Saturn, Uranus, and Neptune and their satellites, rings, and magnetic fields. Filled with figures and photographs. Available to the serious general audience.

Lovett, Laura, Joan Harvath, and Jeff Cuzzi. *Saturn: A New View.* New York: Harry N. Abrams, 2006. A coffee-table book replete with about 150 of the best images returned by the Cassini mission to Saturn. Includes the planet, its many satellites, and the complex ring systems.

Morrison, David. *Voyages to Saturn.* NASA SP-451. Washington, D.C.: Government Printing Office, 1982. The official account of the Saturn missions. The author is an astronomer and was a member of the Voyager Imaging Science Team. An introductory section describes the revival of interest in planetary astronomy, stimulated by the advancement of space technology. Descriptions of the Pioneer 11 and Voyager Saturn encounters are provided. The book ends with a section on Saturn's rings. Well illustrated. Contains a glossary of terms.

Russell, Christopher T. *The Cassini-Huygens Mission: Orbiter Remote Sensing Investigations.* New York: Springer, 2006. Provides a thorough explanation of the remote sensing investigations of the Orbiter and lander. Outlines the scientific objectives of all instruments on the spacecraft. Describes the planned 44 encounters with Titan. Only provides early science return.

Washburn, Mark. *Distant Encounters: The Exploration of Jupiter and Saturn.* New York: Harcourt Brace Jovanovich, 1982. Similar to the book by Cooper but more extensive in scope. Washburn describes the atmosphere among the researchers involved with the Voyager program. Astronomical data are provided, and the budgetary problems that the program faced are discussed. Illustrated with quality photographs.

# SATURN'S SATELLITES

**Categories:** Natural Planetary Satellites; Planets and Planetology; The Saturnian System

*Saturn has a remarkably diverse set of satellites. They include gigantic Titan, which retains a thick atmosphere; Enceladus, possessing a vastly reworked surface that includes active geysers; Hyperion, a disk-shaped satellite whose rotation is erratic; Phoebe, moving in a retrograde orbit; and a coorbiting pair called Janus and Epimetheus, to name some of the most interesting of the sixty confirmed satellites.*

### Overview

Prior to the space age, Saturn was known as the beautiful ringed world of the solar system. Many of its numerous larger satellites were discovered prior to the time of interplanetary spacecraft, the most notable being Titan, Saturn's largest satellite and the only one known from telescopic observation to maintain a thick atmosphere. Prior to the Voyager flybys, planetary scientists expected all of the other Saturn satellites to be relatively uninteresting ice inactive worlds. Only Iapetus was a curiosity, since it displayed a very reflective side and an extremely dark side as well. The Voyager flyby results and the Cassini orbiter images and observations revealed Saturn's system to be a miniature solar system in its own right with a variety of extremely interesting and diverse satellites.

When Voyager 1 passed by Saturn's largest satellite Titan in November, 1980, scientists were somewhat disappointed with imagery transmitted back to Earth. Titan appeared as a uniform orange sphere whose outline was blurred by a dense cloud cover. Closer examination found a higher layer of ultraviolet haze. The southern hemisphere has a slightly darker cast than the northern hemisphere. A clear equatorial boundary was noted, and a darker polar ring is evident in some photographs from Voyager 2. Beneath those clouds, Titan proved more interesting. Voyager 1's close passage behind the disk of Titan allowed the use of its radio transmissions to probe the satellite's atmosphere. The pressure at ground level is 1.5 times that of Earth. If Titan's lower surface gravity is taken into account, the implication is that every square meter of Titan has ten times as much gas above its surface as Earth does.

Methane was spectroscopically detected from Earth, but the prime component of Titan's atmosphere proved to be nitrogen. It was suspected that as much as 10 percent of the atmosphere is argon, and methane makes up between 1 and 6 percent of the rest of the atmosphere, increasing in concentration near Titan's surface.

At higher altitudes, solar ultraviolet rays break methane down, and new molecules form as some hydrogen is lost. Spectroscopic observations show traces of hydrogen, ethane, propane, ethylene, diacetylene, hydrogen cyanide, carbon monoxide, and carbon dioxide. Together, these components form the petrochemical smog that so frustrated the Voyager imaging team.

The upper optical haze layer lies about 280 kilometers above the surface. The main cloud deck is about 200 kilometers from the surface. Titan's solid surface is 400 kilometers smaller in diameter than previously thought, smaller than both Ganymede and Callisto in the Jupiter system. Why do these Jovian satellites not have atmospheres? Titan orbits at a greater distance from Saturn than either of these satellites do from Jupiter, so its tidal stress is less. Furthermore, Saturn is twice as far from the Sun as is Jupiter, so solar radiation intensity at Titan is four times weaker than in the Jovian system.

Beneath those tantalizing orange clouds, the surface temperature is only 94 kelvins; combined with the fact that Titan's atmospheric pressure is 1.5 bars, this temperature suggested the possibility of an ethane and/or methane sea on Titan's surface. If tidal stresses heat the interior enough, there may even be icy geysers. The possibility of life arising at such low temperatures appears unlikely, but certainly the carbon chemistry on Titan must be very interesting.

*Mimas, with its huge crater, Herschel.* (NASA/JPL/Space Science Institute)

probe was also outfitted with a gas chromatograph mass spectrometer to determine atmospheric composition. Titan's atmosphere proved to be hazier than expected, as dust particle concentration was greater than previously believed. Wind data suggested that Titan's atmosphere circulated gas from the south to north pole and back again in periodic fashion.

Winds would play a large role in planetary dynamics for this complex world. Indeed, two years after the probe's several hours of data were collected, Cassini scientists came to the conclusion that Titan's crust moves on a subsurface ocean with crust movements in part driven by wind actions. That movement was noted by comparing radar data from the Orbiter taken at different times during the mission in concert with available Huygens data. The proposed liquid subsurface layer would be located 50 to 100 kilometers beneath the crust and include liquid ammonia in a water ice slush. Floating on this layer, the crust was seen to move as much as 30 kilometers over the course of several years of Cassini observations.

The European Space Agency (ESA) provided the Huygens probe, a combinations atmospheric entry probe and soft lander, for National Aeronautics and Space Administration's (NASA's) Cassini program. The Orbiter carried Huygens from launch in October, 1997, to release on Christmas, 2004. For the next two weeks the probe flew independent from the Cassini orbiter, and then it entered Titan's atmosphere on January 14, 2005, and dropped down under a large parachute to a safe touchdown near Titan's Xanadu region. Some researchers expected Huygens to splash down in a cryogenic sea of liquid hydrocarbons. It became clear rather quickly that Huygens had "plopped" down in what some referred to as Titanian mud. Evidence of liquid action on the surface was found, but the original idea of liquid hydrocarbon seas were dashed. Analysis of data sent from the probe on the way down to impact revealed several layers in the atmosphere, most notably a thick haze between 18 and 20 kilometers above the surface. An Aerosol Collector and Pyrolyzer collected samples at different altitudes to determine the pressure of volatiles and organic materials. The

Further examples of Huygens and Cassini images eventually found the evidence proving the existence of ancient shorelines and the presence of liquid on the surface of Titan. Computer models of this dynamic world had to be greatly altered due to Huygens and Cassini data, and many new questions were raised to give Titan an even more mysterious nature. However, it still was believed to be a world rather similar in some ways to an early Earth, just a world frozen at an early point of physical and chemical evolution prior to the development of life. Down in the subsurface liquid layer higher temperatures could permit complex biochemistry, but there was no information produced by Cassini to investigate that supposition.

By measuring gravitational perturbations on the Voyager spacecraft as they flew through the Saturnian system, scientists at the Jet Propulsion Laboratory (JPL) could determine the masses and densities of Saturn's middle-sized satellites. Rhea's bulk density of 1.3 grams per cubic centimeter suggests that it contains more ice and fewer silicates than Titan. It is worth noting that the density of bodies depends not only on their composition but also on how tightly packed they are. The greater the mass, the higher the gravity, and thus the greater the density. Thus Titan, Ganymede, and Callisto, which all approximate the size of the planet Mercury, have very similar densities, about 1.9 grams per cubic centimeter, and the smaller, icy satellites are less dense, even though the composition may be quite similar to that of larger satellites.

Rhea and all the smaller Saturnian satellites lack atmospheres and show some signs of older, cratered surfaces. The trailing side of Rhea is covered with pale, wispy streaks, a type of feature it shares with Dione. These streaks may be evidence of venting of water vapor from these satellites' interiors, perhaps from tidally induced volcanism in the past. On Earth's moon, such activity was on the side facing Earth rather than on the trailing side. Perhaps these wisps were once found on Rhea's leading side but were eventually eroded, much as meteoric dust erases all but the youngest ray patterns around lunar craters.

During a Cassini flyby of Rhea in November, 2005, some surprising results were obtained. While the spacecraft's magnetometer did not pick up any interactions with Saturn's magnetosphere that would have indicated even a meager atmosphere about Rhea, there was evidence of a broad debris disk and one structured ring about this satellite. The debris disk extended several thousand kilometers out from Rhea, hence it was several Rhea radii in expanse. Computer simulations of the gravitational interactions of Saturn and Rhea indicated that this ring could exist for a considerable time. The source for the ring particles about the small satellite could have been a large impact event with Rhea.

Slightly smaller than Rhea, Dione is bit more dense (1.4 grams per cubic centimeter). Its wispy patterns are more marked than those of Rhea, and it also has some long cracks and large areas of fairly fresh ice, without large craters in evidence. Ice may have flowed through cracks during the cooling phase of the satellite. Unlike most substances, water expands when it freezes. Therefore, satellites made mainly of ice, or differentiated with rocky cores and mantles composed largely of water, might show such expansion cracks. Such cracks are evident on Tethys, and were noted by Voyager 2 on the Uranian moons Ariel and Titania in January, 1986.

Tethys is similar in size to Dione, but it features one huge impact crater. The crater's floor is quite flat, suggesting internal flooding resulting from impact heating. Running from the crater three-quarters of the way around

Tethys is a single gigantic valley system, Ithaca Chasma. Tethys's craters are lower in relief than lunar craters. The icy crust of these Saturnian satellites is more plastic than is lunar crust. Older terrain cratering appears to have been just as heavy as on Earth's moon, resulting craters appear less rugged than the lunar highlands.

The small satellite Mimas features a notable exception to the above rule. One huge crater, Herschel, is one-third as large as Mimas itself. This crater is very deep, about nine kilometers, with a central peak about six kilometers high. It constitutes one of the most striking geological features in the solar system. This crater on such a small roughly spherical body gives it the appearance of the Death Star station in the movie *Star Wars*. Many of the planetary scientists on the Voyager and Cassini teams, and astronomy professors world wide, fondly refer to Mimas as the "Death Star Moon."

It is likely that an impact of any greater force would have broken Mimas apart. Such huge impacts might also explain the very jumbled appearance of Uranus's Miranda, which was apparently broken into several large pieces, then haphazardly reassembled by gravity later.

One highly speculative hypothesis may account for such massive impacts and the intense cratering that is evident throughout the solar system. Perhaps a terrestrial planet in an unstable orbit beyond Mars and an outer icy Jovian satellite were totally fragmented in a high-energy head-on collision. The lighter, icy debris might account for some of the comets. Heavier chunks may have found a relatively stable orbit and formed the asteroid belt. Many chunks and particles, however, would have been scattered in all directions to impact other worlds. In some cases, the impact would have been forceful enough to send up debris which, in turn, would bombard neighboring worlds. This scenario might explain ices found on certain asteroids, the existence of captured satellites such as Phoebe and Phobos, and the fact that meteoritic material seems to have originated on a differentiated planet. It might also explain how fragments which scientists agree came from lunar basalts and from Mars are found on Earth.

While almost a twin of Mimas in size, and orbiting just beyond it in the satellite system, Enceladus is a very different world up close. Even the Pioneer 11 data indicated that it has an albedo near 100 percent. It seems to be made of fresher ice, reflecting far more sunlight than most Saturnian satellites. Had its material been older, dark meteoritic and cometary dust would have darkened it. Voyager's cameras revealed that one of its hemispheres is heavily cratered and fairly old, but the opposite side features smooth plains cut by grooved terrain, similar to Ganymede. This evidence of much rifting and recent internal activity appears on a satellite about one-tenth the size of Ganymede. A count of ring particles in Saturn's extended E ring also found that they peaked near Enceladus. Just as the dust ring of Jupiter is supplied by Io's volcanoes, icy geysers on Enceladus periodically shoot debris above this active world. Why is this small world active at all? Mimas lies closer to Saturn and thus is more tidally stressed, yet it shows no such activity. Nor does the proximity of any other large satellite seem to account for the heating required to generate such activity. Such activity on smaller satellites is not unique; Uranus's moon, Ariel, which is similar in size to Mimas, shows obvious broad rift valleys. The source of heating for this extensive and possibly continuing crustal activity is a mystery.

Cassini found geysers on Enceladus near its south pole along long cracks which essentially act like vents. Fresh crystalline ice forms at the site of these cracks and colors the features distinctively. Cassini scientists dubbed these nearly parallel cracks found at the south polar region "Tiger Stripes." The Tiger Stripes were found to be 124 kilometers long and 40 kilometers apart. This activity

was not new, so why did the Voyagers fail to see these geysers? Voyager 2 flew over Enceladus's north pole and missed them. Cassini's near-infrared mapping spectrometer and solid state imager both examined the ice around the Tiger Stripes. Freshly formed ice was crystalline. As time progresses that pristine ice becomes radiation-damaged amorphous ice.

Data and the geyser actions strongly suggested the presence of a subsurface ocean on Enceladus. However, calculations about heat transport inside Enceladus led researchers to believe that the satellite's subsurface ocean would not be able to exist for more than 30 million years if it were warmed only by heat escaping from the core toward the crust. Since that ocean most likely has been in existence for more than 30 million years, the heating mechanism for both the subsurface and the cryovolcanic activity at the satellite's south pole must be from tidal flexing. Only that could provide the 5.8 gigawatts of heat Cassini saw emerging from the Tiger Stripes over which it flew on a close fly. Since internal heat sources apart from tidal flexing produce only 0.32 gigawatts, without tidal heating Enceladus's subsurface ocean would have frozen. But the story here is more complex than that. Without the subsurface ocean, tidal flexing of the magnitude

*Saturn's second largest moon, Rhea, may have a debris ring of its own, as illustrated in this artist's rendition.* (NASA/JPL/JHUAPL)

necessary to produce the observed heat would not be possible, and without the heat for the tidal flexing the ocean would freeze.

Iapetus confronts scientists with another mystery. A portion of this satellite is extremely dark, whereas the rest of Iapetus has an albedo typical of an icy surface. Iapetus's dark side is six times lower in albedo than that icy portion. Within the darker portion is an irregular dark spot. This pattern of darker leading hemispheres is also seen on Rhea and Dione, but to a far lesser degree. The brighter, icy side does have an albedo of 50 percent. This is typical of older water-ice crusts, and it shows heavy cratering typical of other similar satellites. Some larger craters near the boundary between the hemispheres have light-colored walls, with darker flat floors, like some larger craters on Earth's moon. What is the reddish-black material that gives the darker side an albedo of only 5 percent? It is probably an organic tar, and it appears to be a good match with carbonaceous chondrite meteorites, the dark rings of Uranus, and the black crust of Halley's comet. Carbon almost always appears in an oxidized form (carbon dioxide, carbonic acid, carbonate rocks, carbohydrates) in the inner solar system. Dark neutral carbon is a major solid material found in the outer solar system. It was impossible to judge the age of Iapetus' dark spot, as Voyager's cameras could not pick up any details on the dark side. In fact, some photographs even make the dark side disappear into the blackness of space. The concentration of the dark material on the leading side suggested an external source for this coating. It had been hypothesized that dark Phoebe was responsible. Phoebe's color, however, does not match the black side of Iapetus. The dark floor of some of Iapetus's craters constitutes evidence for an internal origin. New data and insights would have to wait for Cassini to pass by Iapetus at a much closer distance than had the Voyagers.

Cassini flew within 1640 kilometers of Iapetus on September 10, 2007. Unfortunately during the encounter, Cassini entered a safe mode after on-board delicate solid state electronics suffered a cosmic ray hit. Fortunately, most of the science harvest was recovered after a short delay in playback. Among the findings was a raised area around the satellite's mid-section that gave Iapetus the appearance of a walnut. Why the equatorial bulge on this unusual satellite? Julie Castillo of the Jet Propulsion Laboratory advanced an innovative explanation for the unique feature. Castillo invoked a high rotational speed early in the satellite's history coupled with heat from internal radioactivity, perhaps from aluminum 26 and iron

60 isotopes, that softened the satellite to form the equatorial bulge. Consideration of the time frame in which tidal forces forced the spin rate to diminish led to the conclusion that this particular pair of isotopes would be required, since they would be abundant and would generate heat quickly due to rapid radioactive decay. Then, from a softened and malleable state, the satellite's bulge was frozen in place before Iapetus's spin rate slowed down. More investigation would be needed to confirm or refute this theory. Unless extended mission priorities are changed and trajectories reevaluated, this could be the closest Cassini would ever get to this highly unusual satellite.

Saturn's outermost known satellite Phoebe may be a captured asteroid from the far reaches of the main belt, and therefore similar to Chiron. Phoebe's orbit is retrograde, like those of Jupiter's four outermost moonlets. All these small worlds are quite distant from the gas giants whose gravity trapped them. Cassini encountered Phoebe on the way into Saturn orbit insertion. Photographs revealed Phoebe's surface to look almost spongelike, not at all like the other Saturnian satellites.

Like the dark side of Iapetus, Phoebe has an albedo of about 5 percent, which is similar to those of two other captured asteroids, Deimos and Phobos, which orbit Mars. At 200 kilometers in diameter, Phoebe is much rounder than the odd "hamburger moon," Hyperion, which orbits between Tethys and Iapetus. Puck, a satellite of Uranus, is similarly round and dark, and about the same size as Phoebe. This fact suggests that a round shape is the norm for dark, primitive bodies such as these, and that something unusual happened with Hyperion.

Hyperion's shape is quite striking. It is a huge disk, about 250 kilometers across but only 150 kilometers thick. Like Phoebe, its surface is dark, old, and heavily cratered. Stranger still is Hyperion's rotation period. It has not yet been well defined. Like Earth's moon, Saturn's other satellites are tidally locked, with one side permanently facing the planet. As the Voyager 2 team tried to orient photographs of Hyperion to map it, however, they found that it was rotating chaotically. It appears to have no regular rotation period; it tumbles irregularly. Close coupling between Hyperion's eccentric orbit and that of Titan may cause this unique effect.

The nine satellites discussed to this point were known well before the Voyager missions, but Voyager photographs found or confirmed eight more satellites, making Saturn's the most numerous satellite system. Several photographs suggested the existence of even more Saturnian satellites, but their periods of revolution and orbits had

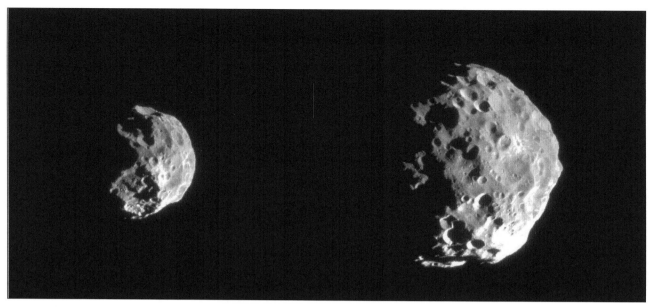

*Phoebe, outermost known satellite, with its heavily cratered surface, may be a captured asteroid. Two images of the satellite shown here were imaged from the Sun-Phoebe spacecraft.* (NASA/JPL/Space Science Institute)

to be determined before they were formally recognized. All these new satellites are much closer to Saturn than are Phoebe and Iapetus, and they show a much more reflective, icy surface like Enceladus. None of these satellites is large enough to be nearly spherical or to have become differentiated. All of them have quite interesting orbits.

Just as the Trojan asteroids share Jupiter's orbit, so two of Saturn's middle-sized satellites have smaller companions in their orbits. Dione has two; Helene, the leading one, appears quite elongated, while the following one appears rounder. Lagrangian, Tethys's companion, is a smaller version of Mimas, with a huge crater from an impact that almost destroyed it.

Saturn's coorbital satellites were first spotted in 1966, but for more than a decade thereafter they were mistaken for a single satellite with an orbit under that of Mimas. Even prior to the Voyager flights, however, observers repeatedly noticed inconsistencies that led some to argue that there must be two satellites sharing the same orbit. Janus, the larger, is about 200 kilometers across, and Epimetheus is about 150 kilometers across. Actually, their orbits are not quite identical. The inner satellite has a period of 16.664 hours; the outer one has a period of 16.672 hours, or a difference of 29 seconds per orbit. Every four years, the inner satellite overtakes the outer at the speed of nine meters per second, and they exchange orbits. This

close relationship and the irregular, elongated appearances of these satellites suggest they were once part of a single larger one split apart by a collision into the two pieces now sharing the same orbit.

The inner three satellites discovered by Voyagers 1 and 2 are all closely associated with Saturn's rings. Atlas, a tiny, football-shaped body, orbits just outside the bright A ring of Saturn. Prometheus is a shepherding moon, keeping the particles in Saturn's F ring in place from the inside of that ring. Pandora plays a similar role on the outside of the F ring. Their close relationship to this thin set of ringlets may explain why the F ring sometimes appears braided. Additional satellites were identified in subsequent reviews of Voyager and other available data. Then, with the arrival of the Cassini probe in the Saturn system, the number of recognized satellites again increased significantly, reaching 60 by 2008.

**Knowledge Gained**

While Jupiter possesses four satellites comparable in size to Earth's moon or even to Mercury, Saturn has only one, Titan. Like Jupiter's Ganymede and Callisto, Titan is comparable to Mercury in size but only about one-third as massive and dense. Its exact dimensions were still in debate prior to the Voyager missions. Its visible orange disk made it appear to be the largest known natural satellite;

out-of-date astronomy textbooks will list Titan as the largest satellite in the solar system. Spectroscopic observations plainly revealed an atmosphere with gaseous methane and other hydrocarbons. Just how deep was the atmosphere, and what was it made of? These questions led the Voyager 1 team to target Titan as a main mission objective and to guide one probe closer to this satellite than to any other body on its mission.

The chief discoveries that resulted concerned Titan's atmosphere. It is thick, twice as dense as Earth's, but like Earth's, Titan's atmosphere is made primarily of nitrogen. Orange clouds appear to be a hydrocarbon smog, with complex organic chemistry taking place there. Surface temperatures and pressures lie close to the triple point of methane, so the surface might experience methane rains that would build up into lakes of liquid methane and freeze into methane ice at the poles. Confirmation of that would have to wait for a probe outfitted with imaging radar and/or a lander. Thus the origin of the Cassini mission. Combination of radar images taken from orbit with data from the Huygens lander eventually confirmed the presence of cryogenic lakes and found ancient shorelines of lakes no longer existent. Huygens appeared to have landed in a wet slushlike material at cryogenic temperatures rather than floating on a lake or sea or even having hit a hard icy surface.

By the time its primary mission was completed, Cassini had flown past Titan several dozen times at varying distances. Perhaps one of Cassini's biggest surprises was the detection that the surface of Titan moved as much as 30 kilometers between the earliest flyby of the Cassini primary mission (2004) and some near the time that the extended mission was approved (2008). This suggested that the crust floated on a layer of fluid, meaning the large satellite likely has an underground ocean, presumably a mixture of water and ammonia.

All of Saturn's remaining satellites are smaller than Earth's moon. Rhea is next in size, about half as large as the Moon at 1,500 kilometers in diameter; Voyager 1 showed its icy surface to be cratered, but with fresher ice creating wispy terrain. Dione is next in size, at 1,100 kilometers in diameter, and has even more wispy terrain than Rhea. Tethys featured a huge, flattened crater on one side, with a great crack or rift running to the other side.

The innermost of the satellites well-known prior to the Voyager flybys are Mimas and Enceladus, both about 500 kilometers in diameter. Mimas was found by Voyager 1 to have a dramatic impact crater one-third as large as the satellite. Enceladus is one of the most puzzling satellites,

with tidal stresses producing plate activity, according to Voyager 2 data. These satellites are, in order from Titan inward: Dione, Tethys, Rhea, Enceladus, and Mimas. Prior to the Voyagers, only their orbital periods and approximate diameters were known, based on their brightness. No one had actually seen their disks. Cassini provided data that made Enceladus much more interesting to planetary scientists.

The brightness of Iapetus presented a major problem. When Gian Domenico Cassini found it in 1671, he realized that this odd satellite must be far brighter on one side (the leading hemisphere as it orbits Saturn) than on the other. Diameter measurements were impossible until Voyager photographed the disk. It proved to be about half as big as the Moon, with one side bright and icy. The other side is mostly covered with a layer of tarlike black material that hid any surface features from the cameras on Voyager 2. Similarly, little was known about Hyperion, another dark satellite orbiting between Titan and Iapetus; it was found by Voyager 2 to be irregularly shaped and tumbling without any rotational period.

Phoebe, the outermost known satellite, is distinguished by its retrograde orbit, like four of the outermost satellites of Jupiter. Like the dark side of Iapetus, Phoebe may be covered with carbon-rich material. More puzzling, the existence of Janus, a tenth moon, had been suspected, but before Voyager, photographs showed it in the wrong place. Voyager 1 detected two satellites sharing the same orbit. The other seven of Saturn's major satellites were not known prior to the Voyager missions. Cassini added considerably to the total list of Saturn's family of satellites.

## Context

Practically nothing was known about Saturn's satellites prior to the Voyager flybys. Titan, Dione, Mimas, and Rhea were examined most fully by Voyager 1, in November, 1980. Until the arrival of Cassini in Saturn orbit, most information about Enceladus, Iapetus, Hyperion, and Tethys had come from Voyager 2 in August, 1981. Much about these satellites were discovered or confirmed by Voyager 1, but thanks to improved orbital data, they were best photographed by Voyager 2. Clearly, a strong argument can be made for using two spacecraft in flyby missions.

In brief, the Voyager missions found Saturn's satellite family to be a very diverse lot. Even satellites similar in size and mass, such as Mimas and Enceladus, appeared very different up close, and obviously were shaped by different processes. Each satellite has its own history of impacts. Tidal stress has played an important role in the evolution of many of these bodies, as it has in the Jovian satellite system. Each satellite has its own fascinating evolutionary story to be interpreted by geologists.

With Cassini repeatedly orbiting Saturn and conducting numerous flybys of many of the satellites, planetary scientists were able to make comparisons over time. Just as the Voyagers had piqued interest in satellites that had once been thought to be merely crater-pocked ice balls, Cassini images revealed many of the satellites not well studied by the Voyagers to also be rather intriguing in totally unexpected ways. Interest in Endeladus, for example, increased greatly due to Cassini observations.

*J. Wayne Wooten and David G. Fisher*

## Further Reading

Consolmagno, Guy. *Worlds Apart: A Textbook in Planetary Sciences.* Englewood Cliffs, N.J.: Prentice Hall, 1994. A text accessible to college-level science and nonscience readers alike. Presents subjects at low-level mathematics and also involves integral calculus where required. Demonstrates how the area of planetary science progresses by questioning previous understanding in light of new observations.

Encrenaz, Thérèse, et al. *The Solar System.* New York: Springer, 2004. A thorough exploration of the solar system from early telescopic observations through the space missions that have investigated all planets with the exception of Pluto by the publication date. Takes an astrophysical approach to give our solar system a wider context as just one member of similar systems throughout the universe.

Harland, David M. *Cassini at Saturn: Huygens Results.* New York: Springer, 2007. This text provides a thorough explanation of the entire Cassini program, including the Huygens landing on Saturn's largest satellite. Essentially a complete collection of NASA releases from the start of Cassini flight operations through the majority of Cassini's seventy orbits of its primary mission. Cassini's primary mission concluded a year after this book entered print. Technical writing style but accessible to a wide audience.

_____. *Mission to Saturn: Cassini and the Huygens Probe.* New York: Springer Praxis, 2002. Another book in Springer's Space Exploration Series, this is a technical description of the Cassini program, its science goals and the instruments used to accomplishment those goals. Written before Cassini ar-

rived at Saturn. Provides a historical review of pre-Cassini knowledge of the Saturn system.

Hartmann, William K. *Moons and Planets.* 5th ed. Belmont, Calif.: Thomson Brooks/Cole, 2005. An updated version of a classic text on planetary science. The chapter on Saturn covers all aspects of ground-based and spacecraft observations of Saturn.

Irwin, Patrick G. J. *Giant Planets of Our Solar System: An Introduction.* 2d ed. New York: Springer, 2006. Suitable as a textbook for upper-level college courses in planetary science. Focuses on Jupiter, Saturn, Uranus, and Neptune and their satellites, rings, and magnetic fields. Filled with figures and photographs. Available to the serious general audience.

Leverington, David. *Babylon to Voyager and Beyond: A History of Planetary Astronomy.* New York: Cambridge University Press, 2003. An historical approach to planetary science. Heavily illustrated, concludes with a summary of spacecraft discoveries. Suitable for general readers and the astronomy community.

Lorenz, Ralph, and Jacqueline Mitton. *Lifting Titan's Veil: Exploring the Giant Moon of Saturn.* Cambridge, England: Cambridge University Press, 2002. An in-depth examination of all that was known about Titan prior to the Cassini-Huygens mission written by an engineer who worked for the European Space Agency and an astrophysicist who was Press Officer for the Royal Astronomical Society. Describes the mission of Cassini-Huygens, but the book was published before the spacecraft arrived at Saturn.

Morrison, David, and Tobias Owen. *The Planetary System.* 3d ed. San Francisco: Pearson/Addison-Wesley, 2003. A fine survey of the solar system, and very current for the editing date. Intended for use as a introductory text, it is very well organized. Highly recommended for the general reader.

Van Pelt, Michel. *Space Invaders: How Robotic Spacecraft Explore the Solar System.* New York: Springer, 2006. An historical account of robotic planetary science missions attempted by all spacefaring nations written by an European Space Agency cost and systems engineer. As such the narrative not only explains the science but also provides a behind-the-scenes description of the development of a space exploration mission from concept proposal to flight operation.

# TITAN

**Categories:** Natural Planetary Satellites; The Saturnian System

*Saturn's largest satellite, Titan, is the only satellite in the solar system with a thick atmosphere. Astronomical observations made from Earth established some time ago that Titan has a density slightly greater than compressed ice, indicating a composition primarily of ice but with a relatively small rocky core. Observations made by the Cassini-Huygens spacecraft show a surface with multiple hydrocarbon lakes that could be breeding grounds for primitive living organisms.*

## Overview

Titan is Saturn's largest satellite, with a diameter of about 5,150 kilometers. Its atmosphere (whose density is several times that of Earth's atmosphere) was discovered in 1944 from spectral analyses of sunlight reflected from the cloud cover. The spectral data indicated the presence of methane gas ($CH_4$). Additional Earth-based observations in 1973 showed a reddish, hazy atmosphere, which was assumed to be photochemical smog created by ultraviolet light from the Sun acting on the methane and other hydrocarbon compounds.

Because Titan is such an unusual satellite, the Pioneer 11 spacecraft flew by Titan in 1979, followed by Voyager 1 and 2 in 1980 and 1981, respectively. Unfortunately, their instruments were not sensitive enough to penetrate Titan's thick atmosphere, although it was learned that its major constituent is nitrogen, with methane and smog making up less than 10 percent of the atmosphere. Hazy smog is formed as the methane is catalyzed by ultraviolet light from the Sun to form more complex organic molecules, similar to the manner in which photochemical smog is produced from unburned fuel in the exhaust emitted by vehicles in Earth's large cities. Because Titan is quite far from the Sun and relatively little of the available light would penetrate the thick clouds, the surface temperature was predicted to be about 94 kelvins.

Calculations indicated that at these temperatures methane should condense from the clouds and fall as rain. The denser organic molecules created in the atmosphere, such as ethane ($C_2H_6$), would also eventually settle on the surface as a layer of malodorous slime. It was thus assumed that the icy surface was covered by either an ocean of liquid methane or a hydrocarbon swamp, with frequent rainstorms of methane.

In 1994, scientists used the Hubble Space Telescope at near-infrared wavelengths (where the haze is more transparent) to map some of Titan's surface features according to their reflectivity. Although details were not resolvable, light and dark surface features were recorded over Titan's sixteen-day rotation period, and one bright area the size of Australia was documented. Definitive conclusions about the nature of the dark and bright areas could not be ascertained, but images proved that the surface was not a global ocean of methane and ethane, as had been assumed; at least part of the surface is solid. Although definite

*A composite image of Titan from the Cassini spacecraft.* (NASA/JPL/University of Arizona)

conclusions could not be made, it was thought that the bright areas were major impact craters in the frozen surface. Information gleaned from this research provided important background information for the Cassini mission, the program of National Aeronautics and Space Administration (NASA) that sent a robotic spacecraft to study Saturn and its satellites. In addition to data to be gathered from flybys, Cassini would release the European Space Agency's Huygens probe, which would parachute to the surface. Images of Titan from the Hubble telescope were used to locate an optimum landing site and to predict how Titan's winds would affect the parachute as it descended through the atmosphere.

The Cassini spacecraft was launched in October, 1997, for its seven-year voyage to rendezvous with Saturn. Beginning to orbit Titan in 2004, it flew 1,192 kilometers above the surface, using infrared cameras and radar to produce detailed maps. It detected irregular highlands and smoother dark areas, including one large region, about the size of Lake Ontario (232 by 72 kilometers), so reminiscent of a lake that its perimeter even exhibited sinuous drainage channels leading to an apparent shore-like boundary. Because the surface temperature was so cold (94 kelvins), the lakes were presumed to be liquid methane and ethane fed by streams of dark organic gunk washed by precipitation from the highlands. Methane evaporating from the lakes would replenish the methane in the atmosphere, from which it would eventually precipitate and return to the surface as rain, mimicking the hydrologic cycle on Earth. The fact that this feature appears in Titan's cloudiest region, where presumably storms are intense enough that methane rain reaches the surface, gave credence to the lake hypothesis. Furthermore, Titan's cold temperature would require a long time for liquid methane on the surface to evaporate; thus, a methane-filled lake would remain stable for a long time.

The Huygens probe was released on January 14, 2005. As it descended, it recorded the temperature, pressure, wind speed, and atmospheric composition at regular time intervals. It also radioed back more detailed images of the surface, showing dark drainage networks leading into the smooth areas but relatively few craters, as expected. The bright spots appeared to be "islands" around which dark material had flowed in the past. Other images showed areas evocative of water ice extruded onto the surface and short, stubby, dark channels which could be springs of liquid methane. Although these data suggested running liquids, no clear evidence of liquid methane was detected on the surface. After landing, Huygens probed the surface, which

*The Huygens lander took this image on January 14, 2005, from Titan's surface. The rocks at the front are about 4 to 6 inches in diameter and may be the remains of a lakebed. (ESA/NASA/ JPL/University of Arizona)*

had the consistency of wet sand covered with a thin crust, possibly consisting of ice mixed with small amounts of solid methane. First pictures of the surface showed a plethora of small erosion-rounded pebbles, assumed initially to be rocks or granite-hard ice blocks, on an orange-colored surface. They later were determined to be mixtures of water and hydrocarbon ice. One image pictured tendrils of surface fog, presumed to be ethane or methane.

The concentration of methane in Titan's atmosphere is puzzling, because ultraviolet light from the Sun would dissociate the methane into carbon and hydrogen, which would either react with the nitrogen to form ammonia ($NH_3$) or be dissipated into space. More complex organic molecules, such as ethane, would also be created; being heavier, they would settle to the surface. It has been calculated that atmospheric methane should remain in the atmosphere for fewer than one million years. Consequently, methane must be injected from some surface source. Perhaps it is outgassed from the methane in the icy crust. Another possibility is a methane volcano. Detailed observations have identified one area where ice and methane may be rising to the crust from a subterranean heat source, to form a methane volcanic caldera emitting methane gas. Since Titan is too small to have a molten interior, the heat source driving the release of methane gas is suspected to be tidal heating, the frictional force generated as this massive satellite revolves in its elliptical orbit about Saturn. Several dark surface markings having straight boundaries with preferred orientations suggest the presence of internal tectonic processes.

Analyses of close flybys of Titan made by the Cassini orbiter in 2008 as well as conclusions based on data gathered by flybys made between 2004 and 2006 provided suggestive evidence for the possibility that Titan's surface may have active cryovolcanoes, perhaps spewing water, ammonia, and methane. Cassini recorded variations in brightness and reflectance (the ratio of reflected light to the incident light upon a surface) in two separate regions of Saturn's largest satellite using its Visible and Infrared Mapping Spectrometer. In one region, the reflectance increased significantly and remained at the elevated level; in the other, it rapidly increased and then tailed off again. Both would indicate vapor or liquid being ejected out of an active vent. Such a mechanism would explain why Titan continues to maintain a thick methane atmosphere when, without replenishment, it should have been greatly diminished over the passage of geologic time.

## Knowledge Gained

Using radar, Cassini was still mapping Titan's surface in 2009 and was expected to continue doing so for some

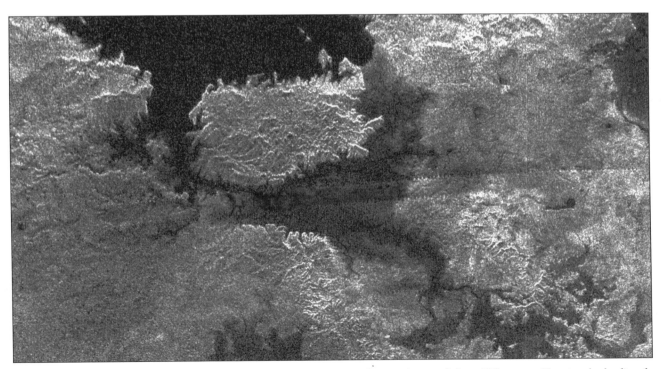

*Cassini captured this image of Titan's surface on February 22, 2007, with a resolution of about 700 meters. The view looks directly down on an island (possibly a peninsula), about the size of the Big Island of Hawaii, in the middle of one of the moon's hydrocarbon lakes.* (NASA/JPL)

*An artist created this image of a lake and smoggy atmosphere of Titan based on data that led scientists to conclude the moon has lakes of liquid hydrocarbons.* (NASA/JPL)

time to come. The hundreds of observed dark areas are believed to be lakes of liquid methane or ethane more than 12.2 meters deep, while shadowy dunes running along the equator are assumed to consist of complex solid organics. Titan's surface seems to contain many gigantic organic chemical factories producing complex hydrocarbons in an abundance surpassing all of Earth's oil reserves. The amount of liquid on Titan's surface is important to ascertain, because methane is a strong greenhouse gas; without atmospheric methane, Titan's surface would be even colder. Liquid methane on the surface could remain, at most, a million years before dissociating and reacting to form heavier hydrocarbon compounds. It is believed that the atmospheric methane is constantly being supplied by volcanic eruptions from the mantle.

In late December, 2008, after much analysis, a group of researchers were ready to publish a scientific article in the research journal *Icarus* reporting the first image taken of a liquid on a planetary surface other than the Earth's. The Mars Phoenix lander had provided clear evidence about six weeks earlier of subsurface water ice at its far north landing site, but when that ice was exposed, it fairly quickly sublimated into the gas phase and became part of the Martian atmosphere. However, an image taken by the Huygens probe after reaching the surface of Saturn's satellite Titan appeared to have clearly recorded a droplet of methane near the edge of the robotic spacecraft itself. The small droplet might have been created by heat emanating from the probe when it condensed humid air to temporarily form liquid methane. In several other images, splotches that appeared and then were not seen in subsequent images of the same area were believed by the authors of the *Icarus* paper also to be droplets of methane.

Titan has an extensive atmosphere, including a methane layer extending 696 kilometers above the surface. There the methane molecules are dissociated by ultraviolet light to form ethane ($C_2H_6$) and acetylene ($C_2H_2$). Cassini images showed two thin haze layers. The outer haze layer, floating about 400 kilometers above the surface, is where additional molecules (such as hydrogen cyanide) are formed from carbon, hydrogen, and nitrogen. About 200 kilometers above the surface, there is a thick global smog of complex organic molecules, produced by chemical reactions among the hydrocarbons dissociated by ultraviolet light. This haze layer absorbs about 90 percent of the incident sunlight, leaving only an orangish haze to reach the surface. It is not currently understood why two separate haze layers are present. Although Titan has a dense atmosphere, it is relatively inefficient at reradiating infrared radiation, thus producing negligible greenhouse warming.

## Context

The Cassini-Huygens mission was a joint venture of NASA, the European Space Agency, and the Italian Space Agency. Enough data were gleaned to keep researchers occupied for years to come.

Titan's surface temperature (94 kelvins) appears to make the satellite a place inhospitable for life to evolve. This environment, although colder, is remarkably similar to that found on Earth billions of years ago, before life began adding oxygen to the atmosphere. Four billion years ago, Earth was covered with warm, shallow seas containing hydrogen, ammonia, and methane gases. From this primordial soup, driven by ultraviolet light and lightning discharges, complex hydrocarbons, including amino acids, formed. Over time, the amino acids linked together to form proteins, eventually creating one that was able to replicate itself. At that point, life was created and molecular evolution became biological evolution.

If life could evolve in Earth's primordial soup, it seems reasonable to suppose that the pools of organic gunk on Titan's surface could form amino acids, if not self-replicating proteins. Studying Titan's prebiotic chemistry can therefore facilitate the understanding of how life may have originated in the universe.

*George R. Plitnik*

## Further Reading

Chaisson, Eric, and Steve McMillan. *Astronomy Today*. 6th ed. New York: Addison-Wesley, 2008. This easily accessible work, written for laymen with inquisitive minds, has an excellent summary of the latest knowledge about Titan as well as pictures from the Huygens landing and an instructive graph of the variation of pressure and temperature as a function of altitude.

Coustenis, Athena, and Fredric W. Taylor. *Titan: Exploring an Earthlike World*. Hackensack, N.J.: World Scientific, 2007. A revised and expanded edition of the 1999 title *Titan: The Earthlike Moon*, this volume summarizes all that is known about Titan through the Cassini-Huygens mission, by two of the project's investigators. Aimed at a general audience, but scientifically rigourous nonetheless.

Hartmann, William K. *Moons and Planets*. 5th ed. Belmont, Calif.: Thomson Brooks/Cole, 2005. This authoritative and regularly updated text considers all the major planetary objects in our solar system. The material is presented by grouping objects under unifying principles, thus elucidating their similarities and their differences as well as the physical processes behind their evolution. Although most of the material is descriptive, some algebra and elementary calculus are included.

Lorenz, Ralph, and Jacqueline Mitton. *Titan Unveiled: Saturn's Mysterious Moon Explored*. Princeton, N.J.: Princeton University Press, 2008. This illustrated tome was the definitive work covering everything known about the surface and atmosphere of Titan at the time of publication. Because Cassini-Huygens was still years away from its encounter with Titan, the authors had to predict—but with some accuracy it turned out—features and conditions on the surface based on limited data.

Sagan, Carl. *Cosmos*. New York: Random House, 1980. Based on the television series of the same name, this lavishly illustrated book includes not only information about Titan's atmosphere but also speculation about the possibility of a methane-based life-form evolving in this environment.

Seeds, Michael A. *Foundations of Astronomy*. 9th ed. Belmont, Calif.: Thomson Brooks/Cole, 2007. This well-illustrated text commingles experimental evidence and theory to provide deep, but well-explained, elucidations of many fascinating facets of the universe. Although only two pages are devoted to Titan, there are four pictures, one of which was taken from the surface by the Huygens probe.

# TRITON

**Categories:** Natural Planetary Satellites; The Neptunian System

*Triton, Neptune's largest satellite, is the solar system's only major satellite that is in a retrograde orbit. It has smoke plumes that astronomers and planetologists cannot explain. It appears to be younger than most of the satellites and planets in the solar system.*

## Overview

The Neptunian satellite Triton is 2,706 kilometers in diameter. The satellite's density is 2.07 grams/centimeter$^3$. With this density, models of the satellite can be constructed. It is thought that there is a metallic core, a silicate mantle, a layer of ice, a possible ocean, and a top layer of ice. The core is expected to have a radius of about 600 kilometers and the mantle a thickness of about 350 kilometers, with a 150-kilometer layer of ice below the ocean and a 250-kilometer layer above the ocean.

William Lassell, a brewer by trade, found Triton on October 10, 1846, using a telescope he built himself. An amateur astronomer, Lassell had been asked by Sir John Herschel to look for satellites of the newly discovered planet, Neptune. Triton is 355,000 kilometers from Neptune. The eccentricity of Triton's orbit is zero, meaning that the orbit is circular. Triton is a most unusual satellite. Its orbit is retrograde, that is, it rotates around Neptune in the opposite direction to the rotation of Neptune. It is synchronous with Neptune, meaning it presents the same face to Neptune at all times. Being synchronous also means that the rotation time of Triton is the same as the time for one orbit around Neptune, 5 days and 21 hours. The angle of inclination is 23°. The angle between the orbit of Triton and the equator of Neptune is 23°.

This large angle of inclination, coupled with the retrograde direction of Triton's orbit, suggests that Triton was captured by Neptune's gravitational field. The method by which this capture occurred is unknown. Triton may have collided with another satellite, causing Triton to slow down enough to be captured and destroying the other satellite at the same time. Another idea is that the other satellite was knocked out of orbit, and Triton was then captured. A third idea is that Triton was part of a binary system. Triton was captured; however, the partner escaped. When Triton was captured, the orbit was probably highly elliptical. The gravitational force of Neptune gradually changed Triton's orbit into the current circular orbit. During the period of this change, the strong pull when Triton was close and the weaker pull as Triton was far away would have caused tidal flexing, or movement within Triton's structure. Such motion would have caused internal friction, generating heat. The heat would have caused differentiation, that is, separation of the components of Triton. Heavier materials, such as metals, would have sunk to the core; medium-mass materials, such as silicates, would have formed a mantle; and lighter materials would have been forced to the surface.

Many believe that Triton is a volcanic satellite because of plumes of dark material that appear to be blown from its surface into the air. This material is concentrated enough to be easily seen. The plumes are about 2 kilometers across, rise as high as 8 kilometers, and consist

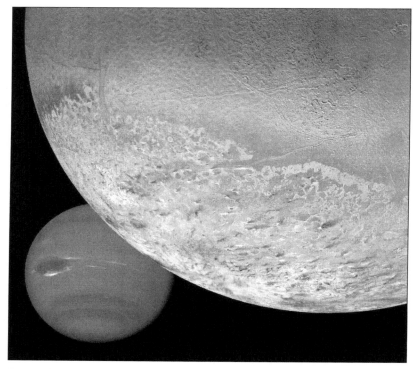

*Triton (foreground) and Neptune appear together in this montage of images from Voyager 2. (NASA/JPL)*

of particles that probably are less than two millimeters in diameter. The size of the particles can be inferred because they do not settle to the surface. These plumes may be putting out 10 kilograms of material per second and may last for years.

Triton has a thin atmosphere, which is composed predominantly of nitrogen at a pressure of 14 millibars (Earth's atmospheric pressure is about 1 bar). There are clouds composed of condensed nitrogen. A diffuse haze can also be seen. The haze probably consists of hydrocarbons and nitriles, produced by the action of sunlight on methane. Wind-driven streaks are oriented in an east-west direction. The wind causes some of the streaks by material blown from the plumes. When all of the streaks, clouds, and plumes are considered, the winds blow northeast close to the surface, eastward at intermediate levels, and westward at the top of the troposphere.

Infrared technology gave scientists the first look at Triton's surface. The spectra could be modeled only by a combination of solid methane ($CH_4$), called methane ice; liquid nitrogen; and water ice. Later spectra showed nitrogen ($N_2$) in a solid or liquid form. One idea is that there is a sea of nitrogen with small amounts of dissolved methane. More likely, there is a layer of solid nitrogen with contaminants of methane, carbon dioxide ($CO_2$), and carbon monoxide (CO). Even at the measured temperature of 38 kelvins, the nitrogen ice will sublime, forming the thin atmosphere first found by Voyager 2. The nitrogen refreezes at the winter pole of Triton, causing a polar ice cap. Solid nitrogen is very transparent; therefore, the sunlight that does reach Triton can heat the interior of the solid nitrogen in a greenhouse effect. Nitrogen, in either gaseous or liquid phase as a result of heating nitrogen originally in solid form, will flow to the surface, where it then freezes. Some of the nitrogen will escape into the atmosphere. The layer of nitrogen thus moves, or at least thins, with the seasons. There is plenty of time for the seasonal shift, because Triton has a 688-year climate cycle due to its unique rotational and orbital motion. Thus, there is a higher albedo (0.7) on the winter end of the satellite,

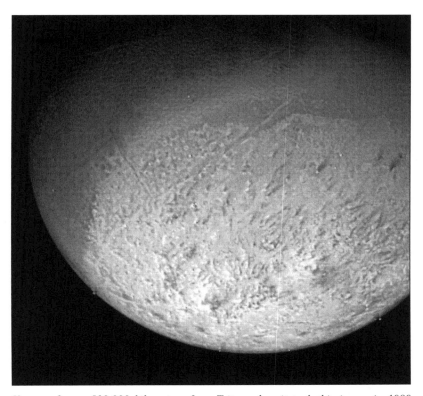

*Voyager 2 was 530,000 kilometers from Triton when it took this image in 1989 through green, violet, and ultraviolet filters.* (NASA)

where the layer of nitrogen ice is thick, and a lower albedo (0.55) on the summer end of Triton. The summer end will show methane ice, which has a reddish color. Radiation will eventually turn the methane dark. The fact that the methane is not dark means that the methane ice is refreshed on a short timescale.

Triton displays at least three different types of surfaces: a bright polar area; areas of dark patches surrounded by lighter material; and high, walled plains. The bright polar area seems to be nitrogen ice on top of a cantaloupe-like, dimpled surface. The dimples, called cavi, are caused not by volcanoes but by diapirism. Diapirism is generated by a gravitational instability in which less dense material flows up through denser material. The density gradient may be caused by temperature or by a difference in composition. The implication on Triton is that the crust has distinct layers, and that the top layer is no more than 20 kilometers thick. The area also has linear ridges across the cantaloupe terrain.

Dark patches within the lighter material are called maculae and probably are composed of carbonaceous material, such as methane ice. Bright material is probably nitrogen ice. Maculae may mark spots of heat that have lost the nitrogen ice layer, allowing the methane ice, with its lower albedo, to show through.

The high plains are caused by a flow of volcanic ice. Some of the plains are smooth plains with a flat-to-undulating structure. Other plains are surrounded by a terraced wall or steppes, called scarps. These plains are very flat, implying that they were filled with liquid at one time. Scarps may be the remainder, as the material on the plains sublimed.

One unique feature of Triton is its small numbere of craters. There is one crater that is 27 kilometers across, named Mazomba, but the small number of craters suggests either that the surface of Triton is very young or that the surface must have been refreshed fairly recently—or both. Part of Triton's surface is considered cryovolcanic instead of silicate-magma volcanic. Cryovolcanic activity is the eruption from the subsurface of icy-cold liquids, which then refreeze on the surface in a more or less smooth structure.

## Knowledge Gained

Much of what has been discovered about Triton has come from Earth-based instruments and the Hubble Space Telescope (HST). Hale Observatory used narrow-band spectrophotometry to determine that Triton had a constant spectral reflectance. Astronomers have compared data from HST, the Infrared Telescope Facility at the University of Hawaii, and Voyager 2 over a period of years to see if there is a seasonal change in Triton's surface. It appears that there is a change, but since the climate cycle is so long, the data remain inconclusive. Both types of surface composition, methane ice and solid nitrogen, were detected by infrared spectra from an Earth-based instrument.

Voyager 2 provided more information in a short time than the land-based instruments had been able to gather in the years since Triton's discovery. The spacecraft's small changes in flight path caused by Triton's mass allowed that mass to be calculated. Pictures allowed the size to be determined. Density could then be calculated. Models of the structure of the satellite could then be developed. The density,

2.07 grams/centimeter$^3$, indicates that there is a large component of silicate materials, even though they do not show in infrared spectra because they are under ice. Voyager also detected the nitrogen atmosphere.

Pictures showed the effects of wind on Triton, a phenomenon that was unexpected. Varied terrain and plumes were also noticed in the pictures. The temperature measured was the coldest of any surface measured. Even with the cold temperature, the different terrains indicated that the surface had been refreshed more recently than any other moon or planet except those planets or moons that are geologically active.

## Context

The density of Triton is close to that of the Pluto-Charon system. Is that a coincidental fact, or are Triton and Pluto related? Could Pluto at one time have been a satellite of Neptune that was knocked off by Triton? The orbital inclination, rotational speed, and retrograde motion all point to some cataclysmic occasion that produced Triton as a satellite.

Triton's surface features raise interesting questions about its energy source. Triton's is the coldest surface in the solar system, yet it also appears to be active, given the plumes and smoothness observed. How can these conditions coexist? The idea that enough sunlight penetrates a deep sheet of solid

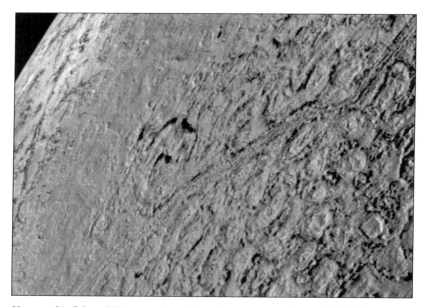

*Voyager 2's flyby of Triton produced this close-up image (from 40,000 kilometers), showing the moon's unique northern hemispheric terrain of more or less regularly spaced circular depressions ringed by ridges. These are not impact craters but more likely areas of collapsed ice.* (NASA/JPL)

nitrogen to produce a greenhouse effect under the ice is startling but may be true; it does appear that the ice sublimes and then refreezes in another place. Yet where does Triton get the energy to produce plumes rising 8 kilometers into the satellite's tenuous atmosphere? Even given the satellite's rather low gravitational acceleration, this phenomenon completes the satellite's overall mystery. Is Triton's interior heated radiogenically? Is there another heat source?

It is possible that scientists do not understand as much about the effect of very low atmospheric pressure, low temperature, and heavy mass as was previously thought, because the planet's heat is generated somewhere. The theory that Triton is heated radiogenically will have to wait until the subsurface can be monitored for radioactive isotopes, or for their daughter isotopes, before it can be confirmed or disproved. Certainly the answer to the question of Triton's energy source will add both to our understanding of the origins of the solar system and to our knowledge of energy physics.

*C. Alton Hassell*

### Further Reading

Bond, Peter. *Distant Worlds: Milestones in Planetary Exploration.* New York: Copernicus Books, 2007. The author discusses each of the planetary systems, including planets, moons, and rings. Exploratory space missions and how they have developed our knowledge of each system are also addressed. Illustrations, bibliography, appendix, index.

Corfield, Richard. *Lives of the Planets.* New York: Basic Books, 2007. The author takes the reader through the different planets and the information gathered by space missions that investigated them. Index.

Croswell, Ken. *Ten Worlds: Everything That Orbits the Sun.* Honesdale, Pa.: Boyds Mills Press, 2007. Basic information on each system is presented separately. Illustrations, bibliography, index. For younger readers.

Cruikshank, Dale P., ed. *Neptune and Triton.* Tucson: University of Arizona Press, 1995. Voyager 2's 1989 encounter with Neptune revealed Triton to be a frozen, icy world with clouds, haze layers, and vertical plumes of particles rising high into the thin atmosphere. Originally presented as papers at a 1992 conference, the chapters in this volume are all by experts on Neptune, its many satellites, and its near-space environment. Until engineers can design propulsion systems for the next mission to the outer solar system, this 1,249-page tome will remain the most authoritative one-volume resource on Neptune and its satellites.

Dasch, Pat. *Icy Worlds of the Solar System.* Cambridge, England: Cambridge University Press, 2004. This book discusses ice, first on Earth, then on other solar-system bodies, including Triton. Illustrations, bibliography, index.

Hartmann, William K., and Ron Miller. *The Grand Tour: A Traveler's Guide to the Solar System.* 3d ed. New York: Workman, 2005. Focusing on the Voyager missions, this volume addresses each major planet and the major moons, including Triton. Includes outstanding illustrations. Illustrations, bibliography, index.

Irwin, Patrick G. J. *Giant Planets of Our Solar System: An Introduction.* 2d ed. New York: Springer, 2006. Suitable as a textbook for upper-level college courses in planetary science. Focuses on Jupiter, Saturn, Uranus, and Neptune and their satellites, rings, and magnetic fields. Filled with figures and photographs. Accessible to the serious general audience.

Lopes, Rosaly M. C., and Michael W. Carroll. *Alien Volcanoes.* Baltimore: Johns Hopkins University Press, 2008. The focus is on volcanism throughout the solar system, including the possibilities on Triton. Illustrations, bibliography, index.

McFadden, Lucy-Ann Adams, Paul Robest Weissman, and T. V. Johnson, eds. *Encyclopedia of the Solar System.* San Diego: Academic Press, 2007. The editors have collected articles written by many experts. It is one of the best surveys of material about the solar system. Illustrations, appendix, index.

# URANUS'S ATMOSPHERE

**Categories:** Planets and Planetology; The Uranian System

*Uranus is the seventh planet from the Sun. It shares much in common with Jupiter and Saturn, but it is also significantly different from the larger Jovian, or "gas giant," planets. Its atmosphere is composed mainly of hydrogen and helium, but its color is governed by selective absorption of light by methane, which is abundant in greater measure in Uranus's atmosphere than in either Jupiter's or Saturn's.*

### Overview

The planet Uranus was the first to be discovered with a telescope. Its existence was declared by Sir William Herschel on March 13, 1781. After several proposed names, the most curious of which was a proposed reference to the King of England, George III, the planet was named Uranus. From mythology Uranus is the father of Saturn and grandfather of Jupiter.

Uranus is the third largest planet in the solar system. With an orbit that varies from 18.4 to 20 astronomical units (AU, or the mean distance from the Earth to the Sun, namely 150 million kilometers), it takes Uranus eighty-four years to complete one revolution about the Sun. Naturally, at this greater distance from the Sun, Uranus receives far less solar radiation than Jupiter and Saturn. Nevertheless, its location suggested to early researchers that the composition and nature of Uranus were similar to those of Jupiter and Saturn.

Superficially, that is true. Uranus is composed largely of hydrogen and helium. However, the atmosphere of Uranus has been determined to be colder than that of Jupiter and Saturn and has a less dynamic structure than the turbulent atmosphere of Jupiter or the pastel banding of Saturn. Uranus's atmospheric temperature can drop to 49 kelvins. In addition to the preponderance of hydrogen

and helium, the atmosphere has a larger amount of ices and hydrocarbon than do the atmospheres of Jupiter and Saturn. Ices include water, ammonia, ammonium hydro-sulfide, and methane. Selective absorption of radiation, in good measure by methane, results in the planet's pale bluish-green appearance.

By compositional abundance, Uranus is 83 percent hydrogen, 15 percent helium, 2.3 percent methane. The total also includes other, low-concentration gases and hydrocarbons and does not add up precisely to 100 percent given significant uncertainties about the abundance of hydrogen and helium. Hydrocarbons that appear only in trace amounts include ethane, acetylene, methyl acetylene, and diacetylene. These and other hydrocarbons are thought to be produced in the upper atmosphere by photolysis of methane under incident solar ultraviolet light. Carbon monoxide and carbon dioxide have also been detected. Like the planet's water vapor, carbon dioxide and carbon monoxide must have been acquired by impacting comets and infalling dust. All totaled, Uranus has a carbon

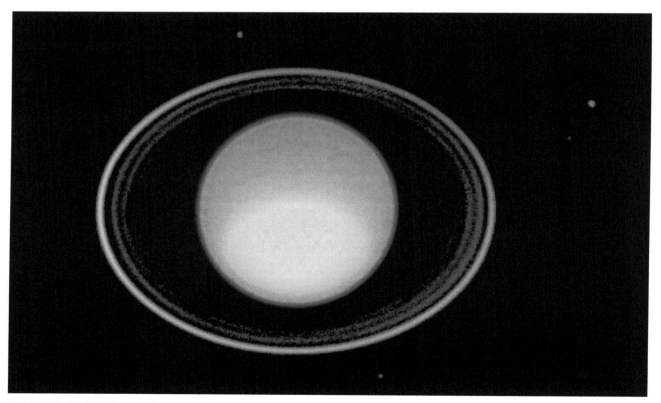

*In 1995, the Hubble Space Telescope revealed much about Uranus's atmosphere: Infrared images showed that it is composed of hydrogen and traces of methane; the inner atmosphere is clear, an intermediate yellow layer is hazy, and a very thin outer layer is red. In addition, Uranus's rings are bright in the infrared. (NASA/Erich Karkoschka, University of Arizona)*

content, primarily found in the atmosphere, somewhere between twenty and thirty times that of solar abundance.

One thing that makes Uranus particularly curious is the fact that its rotational axis is tilted 97.8° from the perpendicular to the ecliptic plane. This tilt is why many refer to Uranus as the planet that rotates on its side. There is no universally accepted explanation for this high degree of tilt, but many believe the planet was knocked on its side by a collision with a large body early in the Uranian system's development. This curious tilt means that for roughly half of each orbit the north pole receives solar radiation, and for roughly half of the rest of the orbit the south pole is in sunlight. This makes for unusual seasons and atmospheric dynamics. Although the planet's interior rotates once every 17 hours 14 minutes, the atmosphere rotates differentially. Features in the upper atmosphere have been clocked at as much as 0.25 kilometer per second and thus may experience a full rotation in less than 14 hours.

## Knowledge Gained

Earth-based telescopic studies of Uranus revealed it to have a bizarre orientation of its rotational axis, several relatively small satellites, and an orbital period of 84.3 years. In 1977, observations made from an aircraft-based telescope as Uranus occulted a star revealed the presence of dark rings around the mysterious seventh planet from the Sun. That same year the Voyager 2 spacecraft launched on an approved mission to fly by both Jupiter and Saturn. The National Aeronautics and Space Administration (NASA) originally proposed sending an armada of sophisticated spacecraft on what had been termed the "Grand Tour." This "tour" referred to the fact that every 176 years, planetary alignments are such that gravitational slingshot maneuvers in the outer solar system can be used to send spacecraft to investigate all the outer planets from Jupiter to Pluto. Unfortunately, that ambitious plan was not funded, but NASA was given authorization to build two modest Voyager spacecraft for exhaustive investigations of Jupiter and Saturn. When Voyager 1 was successful at both gas giants, the Voyager 2 spacecraft was targeted through the Saturn system in such a way as to make possible a flyby of Uranus and Neptune.

Atmospheric structure is often discussed in terms of either pressure levels or temperature or both. If one defines Uranus's "surface" as the site where the pressure is 1 bar (1 Earth atmosphere, or $10^5$ pascals), then that atmosphere can be described as follows. Uranus has a troposphere found from -300 to 50 kilometers above the surface, where the pressure varies from 100 to 0.1 bar,

respectively. A stratosphere exists between 50 and 400 kilometers, where the pressure varies from 0.1 to $10^{-10}$ bar. Then, from 400 kilometers out to as much as two planet radii, or roughly 50,000 kilometers, is the thermosphere and corona, where the pressure dwindles down to near vacuum from the upper stratospheric level of $10^{-10}$ bar.

One might think that because its poles receive more solar illumination than the equatorial region, Uranus would be warmer at the pole presently facing toward the Sun, but that is not the case. Near the equator is the planet's only portion to experience fairly rapid day-night variation due to the excessive tilt of Uranus. Near the equator the warmest temperatures are recorded. Upper atmospheric temperatures near the equator can rise to 57 kelvins. Why the equatorial region is warmer than the illuminated polar region is currently unknown.

The Hubble Space Telescope routinely was used by planetary scientists to examine Uranus for features and changes in those features within the planet's atmosphere. In 1998 an image credited to NASA and Erich Karkoschka of the University of Arizona revealed on the order of twenty clouds in Uranus's atmosphere. That was rather remarkable; prior to that time, in the entire history of Uranus observations, there had been fewer than that number of clouds seen in the planet's usually unremarkable-looking atmosphere. This Hubble image was taken in infrared and clearly showed the planet's rings and many of its known satellites. In addition to the clouds, it revealed a bright band circling the planet. Wind speeds of clouds near the band were determined to be in excess of 500 kilometers per hour. One of the clouds seen in this infrared image was the brightest Uranian cloud ever observed.

In 2006, Hubble images, in concert with near-infrared observations made using the ground-based Keck telescope, revealed a new, more dynamic picture of Uranus. The planet was seen to have a great dark spot and some degree of banding. These observations were summarized by astronomer Heidi Hammel at a Hubble science overview briefing held in 2002, about a month in advance of what was then proposed to be the final shuttle servicing mission to the Hubble. Hammel used her appearance at that briefing as an opportunity to stress how much Hubble had already done to change the picture of an inactive Uranian atmosphere as presented by Voyager 2 in 1986. Hammel and other astronomers had detected an increase in the number and scope of clouds in the ice giant's atmosphere, in addition to finding that great dark spot. She expressed enthusiasm about extended Hubble operations advancing understanding of Uranus, perhaps the planet about which the least is known. This increased activity appears related to seasonal changes as Uranus orbits the Sun.

## Context

Uranus was the first planet for which clear records for discovery exist. Although Herschel was not the first to note Uranus in astronomical records, he was the first to identify it correctly as a planet and not a comet or unidentified star, as others had done previously. From the time of its discovery to the dawn of the space age, little could be learned about Uranus from ground-based telescopes. However, observations of Uranus's orbit about the Sun led to the recognition that there was good reason to believe that it was not the last planet to be discovered in the solar system. Based on gravitational perturbations in the orbit of Uranus, Neptune was discovered by and large by mathematical analysis. Observations verified the correctness of those calculations. Uranus and Neptune were believed to be very similar. Both displayed a bluish-green tint in telescopic views. Spectroscopic analysis indicated that both planets had atmospheres different from those of Jupiter and Saturn. Like their larger gas giant cousins, Uranus and Neptune were known to have atmospheres rich in hydrogen and helium, but their bluish-green color was identified as due to extensive absorption of red light by methane.

Voyager 2 provided the greatest portion of current understanding about the Uranian system. Planetary scientists interested in Uranus await a return mission, most likely an orbiter, perhaps with a lander probe for one of the icy satellites and an atmospheric probe to ram through the upper atmosphere of Uranus and conduct measurements until it is crushed. Ground-based observations and imaging by the Hubble Space Telescope continue in the meantime. The biggest question about detection and observation of clouds in Uranus's atmosphere centers on the source of energy driving those storms, since Uranus's internal heat flow appears to be insufficient to cause such airflow.

*David G. Fisher*

## Further Reading

Elkins-Tanton, Linda T. *Uranus, Neptune, Pluto, and the Outer Solar System.* New York: Chelsea House, 2006. This book explores the Sun's relationship with the three outer planets and their moons, considering these planets as recorders of the formation of the solar system. Aimed at a general or high school audience. Illustrations, bibliography, index.

Encrenaz, Thérèse, et al. *The Solar System.* New York: Springer, 2004. A thorough exploration of the solar system from early telescopic observations through the space missions that had investigated all planets by the publication date. Takes an astrophysical approach to place our solar system in a broad context as just one member of similar systems throughout the universe.

Freedman, Roger A., and William J. Kaufmann III. *Universe.* 8th ed. New York: W. H. Freeman, 2008. A college text on astronomy, somewhat more advanced than many introductory texts but with a wealth of detail and excellent diagrams. Chapters 6 through 16 describe the solar system, including Uranus and what is know about its atmosphere. Comes with a CD-ROM.

Irwin, Patrick G. J. *Giant Planets of Our Solar System: An Introduction.* 2d ed. New York: Springer, 2006. Focuses on Jupiter, Saturn, Uranus, and Neptune and their atmospheres. Suitable as a textbook for upper level college courses in planetary science. Filled with figures and photographs. Available to the serious general audience.

McBride, Neil, and Iain Gilmour, eds. *An Introduction to the Solar System.* Cambridge, England: Cambridge University Press, 2004. A complete description of solar-system astronomy suitable for an introductory college course. Accessible to nonspecialists as well. Filled with supplemental learning aids and solved student exercises.

Miller, Ron. *Uranus and Neptune.* Brookfield, Conn.: Twenty-First Century Books, 2003. Considers Uranus and its satellites in comparison with other gas giants, especially Neptune, including their atmospheres.

Morrison, David, and Tobias Owen. *The Planetary System.* 3d ed. San Francisco: Pearson/Addison-Wesley, 2003. Geared for the undergraduate college student. Planetary atmospheres are treated as important physical features of the various members of the Sun's family. They are discussed individually in the context of what is known about each planet's characteristics and with regard to theories about their evolution and the evolution of the entire solar system. Comprehensive for the average reader.

Tocci, Salvadore. *A Look at Uranus.* New York: Franklin Watts, 2003. As part of the Out of This World series, this book covers all aspects of the planet Uranus from its discovery through 2002. Includes several photographs. Suitable for all readers.

# URANUS'S INTERIOR

**Categories:** Planets and Planetology; The Uranian System

*Uranus, the seventh planet from the Sun, has much in common with the other Jovian gas giants, but its interior is different from the interiors of the larger planets Jupiter and Saturn and more like that of Neptune.*

## Overview

Uranus is largely composed of hydrogen and helium and is considered to be a Jovian planet, or gas giant. The interior of Uranus differs significantly from that of Jupiter and Saturn, however. It shares much more in common with Neptune. Like the interior of Jupiter and Saturn, Uranus's interior cannot be directly sampled. Models of the interiors of the Jovian planets are inferred by external observation. For example, a magnetic field tells much about the nature and physical characteristics of a given planet's interior.

Accurate determinations of orbital motions of Uranus's satellites led to a precise value of the planet's mass. With its size slightly in excess of Neptune, Uranus is the third largest planet in the solar system, but the second least dense. Uranus has a mass 14.5 times that of Earth. With a mean radius four times the Earth's, its overall density is 1.27 grams per cubic centimeter. Of course, the different structures of Uranus, from the core to the upper atmosphere, have specific characteristics of their own. Nevertheless, in gross terms, only Saturn is less dense than Uranus as a planet. Uranus is composed, like Jupiter and Saturn, primarily of hydrogen and helium, although the percentages of both are different in Uranus from the percentages of Jupiter and of Saturn. Uranus contains methane, water vapor, ammonia, carbon monoxide, and carbon dioxide. Hydrocarbons such as ethane, acetylene, methyl acetylene, and diacetylene are also present, presumably created by photolysis of methane in the atmosphere under illumination of solar ultraviolet light.

Like Jupiter and Saturn, hydrogen in Uranus is found in gaseous form in the atmosphere, but deep inside the planet at greater pressure hydrogen may exist in more exotic forms. Whether helium is found in forms other than as a simple gas is a matter of debate. Because of their size, temperature, pressure, compositional, and interior structural differences from Jupiter and Saturn, Uranus and Neptune both are often referred to as ice giants.

Several models exist for Uranus's interior. They agree on major features and differ in only minor ways. A great deal of the interior is composed of water, ammonia, and methane, with the percentage of water included in each model varying. That water percentage ranges between 9.3 to 13.5 Earth masses, depending on which model a scientist supports. With only 0.5 to 1.5 Earth masses of hydrogen and helium in the atmosphere and interior, that leaves between 0.5 to 3.7 Earth masses for what is considered rocky material.

Under the atmosphere of hydrogen, helium, and methane are two inner layers. First, an icy mantle actually incorporates the majority of the planet's mass. Under that is the rocky core. The core represents less than 20 percent of the planet's radius, and the relatively thin upper atmosphere of gases represents another 20 percent of the planet's radius. This leaves the mantle as 60 percent of the planet, with as much as 13.5 Earth masses. Obviously the densities of the atmosphere, mantle, and core are different, just as pressures and temperatures in these three distinctly different regions vary.

The mantle is icy in the sense that it is a hot, dense, electrically conducting mixture of water, ammonia, and less abundant volatile substances. In the mantle, under the great pressures present, molecules dissolved in the water become ionized and therefore create the high electrical conductivity displayed by the mantle. Because of its combined fluid and highly conductive nature, this layer is often referred to as a water-ammonia ocean. The latter quality of the mantle is responsible for generating the planet's complex magnetic field.

The core is far denser, at approximately 9 grams per cubic centimeter. The planet's central pressure and temperature are believed to reach 8 million bars and 5,000 kelvins, respectively. High internal heat flows out to the atmosphere. Materials that have high electrical conductivities usually also have high thermal conductivities. In the case of Uranus, however, it is obvious, based on atmospheric quiescence, that heat flow from the planet's core to the atmosphere is less than that of Neptune. Whereas Neptune radiates more energy than it intercepts by solar irradiance, Uranus is barely 6 percent greater in the infrared than the solar energy absorbed by Uranus's atmosphere. All models agree that Uranus's heat flow from the interior is only 0.042 watt per square meter. The heat flow from the much less massive and smaller Earth, by comparison, is 0.075 watt per square meter. Uranus's low heat flow and hence cold atmosphere (portions of the troposphere have been recorded at a mere 49 kelvins) could be tied to the planet's bizarre rotational configuration. Most planetary scientists believe that Uranus suffered a catastrophic event early in its history, presumably a collision with a large planet-sized body, in order to be left rotating on its side. During such an impact Uranus lost a great deal of its primordial interior heat. Such heat is left over from the gravitational collapse that created the planet. Not all scientists subscribe to that explanation. Some propose that instead the planet's mantle could be layered by composition in such a way that heat flow toward the atmosphere could be diminished by convective action.

## Methods of Study

The Voyager 2 spacecraft has been the only spacecraft to encounter the Uranus system. Originally intended only to fly by Jupiter and Saturn, the Voyager 2 spacecraft was specially targeted to pass through the Uranian system in early 1986. While Voyager 2 cruised to Uranus, a number of astronomers such as Heidi Hammel trained some of the best earthbound telescopes suitable for planetary studies at Uranus to get a feel for what Voyager would encounter. Those studies provided additional information about Uranus's atmosphere but also hinted that Uranus had some internal heat sufficient to drive cloud dynamics. Earthbound telescopic studies did little, however, to enhance understanding of Uranus's interior. A particularly important discovery during the Uranus encounter that aided in describing Uranus's interior was detection of a planetary magnetic field. The nature of Uranus's magnetic field helped planetary scientists devise models for the planet's interior, models quite different from those for the interiors of both Jupiter and Saturn. Then, in 1989, Voyager 2 fulfilled a similar task at Neptune. Planetary scientists realized that the relative similarities of Uranus and Neptune extended to their interiors.

It took the Voyager 2 spacecraft nearly five years to cross the gulf from the Saturn system to the Uranus system. The encounter phase began on November 4, 1985. A great many fundamental questions were about to be answered. At first only ever-increasingly revealing images were produced by Voyager 2. Scientists eagerly awaited detection of a magnetosphere which would indicate Uranus possessed a magnetic field. That would also tell much about the nature of the planet's interior.

When Voyager 2 indeed picked up radio signals indicating the spacecraft crossed the magnetosphere, it clearly revealed Uranus has a magnetic field. After a jam-packed investigation, Voyager 2 ended the Uranus encounter on January 25, 1986, snapping a farewell image of the crescent planet.

## Context

The planets Mercury through Saturn were known to the ancients. No individual can be credited with their discovery. Uranus, however, is the first planet for which definite records exist indicating when the planet was observed and charted. Herschel was not the first to pay special attention to Uranus. Others had misidentified it as an unnamed star or an unknown comet. Herschel recognized Uranus as a newly discovered planet orbiting the Sun. From that time forward to the dawn of the space age, even as Earth-based telescopes grew in size and resolution, little could be learned substantively about Uranus other than its mass, size, mean distance from the Sun, rotation rate, and that it has a very unusual axial tilt relative to the ecliptic plane. It was realized that its atmosphere was quite different from that of both Jupiter and Saturn, suggesting the Uranus's interior differs from those of its two larger gas giant cousins due to its smaller mass. The interior was suspected to be more akin to that of Neptune, which was the next planet in the outer solar system to be discovered after Uranus.

Indeed, models of the two ice giants, generated based on Voyager 2 results and observations, are quite similar. Both planets are believed to have an outer envelope of molecular hydrogen, helium, and methane. Underneath that both planets have mantles that contain water, methane, and ammonia under conditions of high pressures and temperatures. Beneath that is an icy and rocky core. However, the difference between Uranus's and Neptune's interiors is that Uranus's is less active: The planet does not have as great a heat flow from the interior to drive atmospheric dynamics.

*David G. Fisher*

## Further Reading

Burgess, Eric. *Uranus and Neptune: The Distant Giants*. New York: Columbia University Press, 1988. Covers the Voyager 2 spacecraft's mission, technical difficulties, and its encounters with Jupiter, Saturn, and Uranus. Focuses on data collected by Voyager 2 about Uranus. Includes several illustrations and tables. Well written, suitable for the general audience.

Elkins-Tanton, Linda T. *Uranus, Neptune, Pluto, and the Outer Solar System*. New York: Chelsea House, 2006. This book explores the Sun's relationship with the three outer planets and their moons. It looks at these planets as recorders of the formation of the solar system. Aimed at a general or high school audience. Illustrations, bibliography, index.

Encrenaz, Thérèse, et al. *The Solar System*. New York: Springer, 2004. A thorough exploration of the solar system from early telescopic observations through the space missions that had investigated all planets by the publication date. Takes an astrophysical approach to place our solar system in a wider context as just one member of similar systems throughout the universe.

Hunt, Garry E., and Patrick Moore. *Atlas of Uranus*. New York: Cambridge University Press, 1988. This was the first volume after the 1986 Voyager encounter to offer a comprehensive history of Uranus: its discovery, satellites, rings, and the data returned by Voyager, including photographs.

Irwin, Patrick G. J. *Giant Planets of Our Solar System: An Introduction*. 2d ed. New York: Springer, 2006. Suitable as a textbook for upper-level college courses in planetary science. Focuses on Jupiter, Saturn, Uranus, and Neptune and their satellites, rings, and magnetic fields. Filled with figures and photographs.

Loewen, Nancy. *The Sideways Planet: Uranus*. Mankato, Minn.: Picture Window Books, 2008. An educational children's book devoted to the planet Uranus. Covers Uranus's rings, moons, and tilted axis.

Miner, Ellis. *Uranus: The Planet, Rings, and Satellites*. New York: Ellis Horwood, 1990. The author thoroughly covers the topics of both the Uranian system and the Voyager mission. Illustrations, bibliography, index.

Schmude, Richard W. *Uranus, Neptune, and Pluto and How to Observe Them*. New York: Springer, 2008. Ideal for backyard or amateur astronomers who are interested in observing the outer planets. Also includes up-to-date information about the planets.

Tocci, Salvadore. *A Look at Uranus*. New York: Franklin Watts, 2003. As part of the Out of This World series, this book covers all aspects of the planet Uranus from its discovery through 2002. Includes several photographs. Suitable for all readers.

# URANUS'S MAGNETIC FIELD

**Categories:** Planets and Planetology; The Uranian System

*Uranus is the seventh planet outward from the Sun. Although a Jovian or gas giant planet, it has more in common with Neptune than Jupiter or Saturn. Uranus's magnetic field is believed to be produced in a manner similar to that which generates Neptune's magnetic field.*

## Overview

Uranus was the first planet to be discovered through telescopic observations. In March, 1781, while searching for binary stars using a 2-meter telescope, William Herschel noted an object he initially thought was either a comet or a nebula. Recognizing its observed orbital motion to be that of a planet, Herschel first proposed the object be named in honor of King George III of England. However, the planet was finally named after the Roman god of the heavens.

Observations of Uranus continued over the next two centuries, but by and large Uranus remained an enticing mystery. Five satellites were discovered between 1787 and 1948. The rotational axis of the planet proved extremely surprising. Uranus is a world rotating virtually on its side. The axis is inclined 97.8° relative to the ecliptic plane of the solar system. Since Uranus's atmosphere—as seen from Earth-based telescopes between the time of Herschel and the dawn of the space age—did not reveal significant features to observe over time, it was not until the second half of the twentieth century that this planet's rotational period was accurately determined.

Knowledge of the orientation of Uranus's rotational axis, its interior structure, and its rotational rate is key to determining the nature of Uranus's magnetic field. Indeed, until the Voyager 2 spacecraft encountered the planet close up in January, 1986, there was no definitive proof that Uranus even possessed a magnetic field, although one was strongly suspected based on a contemporary model of the planet that shared similarities with gas giants Jupiter and Saturn.

Voyager 2 carried a magnetometer and a radio astronomy experiment used in concert to sample the magnetic environment of each planet (Jupiter, Saturn, Uranus, Neptune) that it encountered on its historic "Grand Tour." Basically, a sophisticated magnetometer operates on the very fundamental principal of an induced voltage being produced in a coil of wire when it intercepts a time-varying magnetic flux. That flux is directly proportional to the instantaneous strength of the magnetic field. Spacecraft such as Voyager 2 detect and investigate magnetic environments in space by picking up radiation in radio wavelengths produced by oscillating charged particles, such as those in the solar wind or ions trapped in planetary magnetic fields.

Uranus and Neptune were considered to be gas giants like Jupiter and Saturn until it was realized, based largely upon computer models and Voyager 2 data, that being smaller in mass and having significant atmospheric differences from Jupiter and Saturn, Uranus and Neptune were better classified as ice giants. Finding out about Uranus's interior would be key to determining the means whereby a magnetic field could be produced by the planet. However, the converse is true as well. Directly measuring the magnetic field of the planet would help to develop a model of Uranus's interior. What was known about Uranus's composition in the atmosphere prior to the Voyager 2 encounter was that the planet is composed primarily of hydrogen (83 percent) and helium (15 percent). However, Uranus has a pale blue-green color due to the presence of methane (2 percent), which selectively absorbs red wavelengths of light. Uranus also contains ices such as water, ammonia, and an assortment of hydrocarbons.

## Knowledge Gained

Voyager 2 determined that the magnetic field generated by Uranus has quite unusual characteristics. The planet's dipole moment is approximately fifty times that of Earth. The value of Uranus's dipole moment is $3.8 \times 10^{17}$ telsa meters cubed. By comparison to giant Jupiter, this is a mere 0.26 percent of the dipole moment of the largest planet in the solar system. Uranus's average magnetic field strength, the maximum value or amplitude of the field, at the plant's "surface" was measured to be 23 micro-teslas. However, magnetic field strength varied with latitude. At the "surface" in the southern hemisphere, the field strength was seen to dip as low as 10 micro-teslas. Then again, at the "surface" in the northern hemisphere the field was found to be as strong at 110 micro-tesla. This situation contrasts greatly with Earth, where the magnetic field is nearly as intense at both poles. Earth's field is centered close to the planet's physical center. However, that is not the case with Uranus. The center of Uranus's magnetic field actually is displaced from the physical center of the planet by about a third of its radius; the magnetic center is closer to the south rotational pole. Further complicating the field is the fact that the magnetic axis, the line from the south to the north pole through the planet, is tilted 59°

relative to the line running between the north and south rotational poles.

Uranus's magnetosphere thus displays a highly unusual tilt. That tilt and the rotational tilt of the planet give the dynamic behavior of Uranus's magnetosphere a twisting structure. To make the magnetosphere's character even stranger, it appears that the ring system around Uranus actually streams ions in the magnetosphere down into Uranus's atmosphere. Auroras are produced, but they differ somewhat from the familiar auroral displays seen in Earth's polar regions.

By detecting variations in radio waves produced by the planet's magnetic field, astronomers have detected the rotational rate of the planet more accurately than by following the differential rotation of those atmospheric features that could be found. That rotation value is 17.233 hours.

Despite the unusual orientation and subsequent twisting behavior of the field lines as the planet rotates, Uranus's magnetosphere does share some things with other planetary magnetic fields in our solar system. Its magnetosphere is affected by the solar wind, forming a bow shock ahead of the planet, a magnetopause, and a magnetotail. The bow shock was crossed at a distance equivalent to 23 Uranus radii, whereas the magnetopause was determined by Voyager 2 to be located at 18 Uranus radii. The magnetotail appears as a corkscrew structure due to the twisting of the planet's magnetic field lines. This magnetic field structure also has given Uranus radiation belts of trapped charged particles. Those particles consist primarily of protons and electrons but have a minor component of molecular hydrogen ions. Uranian satellites create gaps in the radiation belts by "sweeping up" charged particles as they revolve about that planet. Ring particles and the surfaces of the planet's satellites struck by this ionizing radiation are darkened by that exposure.

What generates such an unusual planetary magnetic field? Whereas Earth's field is created deep in the planet by a dynamo effect involving electrical currents generated by its molten core, Uranus's magnetic field is speculated to be produced by the ice giant's mantle. Between the atmosphere and the planet's core, Uranus has a mantle layer believed to be composed of a highly pressurized water, ammonia, and other ices that become ionized under that tremendous pressure and in the presence of temperatures in excess of 1,000 kelvins. Therefore, currents flow through the mantle. Some scientists do not accept this explanation, but at present there is no way, short of direct investigation of the planet's interior, to validate the mantle "ocean" hypothesis or show it to be incorrect.

Why does Uranus's magnetic field have such an unusual orientation relative to the planet's rotational axis? Here there are two reasonable hypotheses. One suggests that the planet is in the process of reversing its magnetic field. (Rocks on Earth present a record that Earth's magnetic field has reversed itself many times over geologic time.) The second hypothesis is that the disruption of the magnetic field alignment resulted from a collision between Uranus and one or more large bodies. Further study will be needed to decide between these two theories or replace them with a better explanation.

The Hubble Space Telescope in concert with the Keck telescope has imaged new storms in Uranus's atmosphere. If Hubble received another servicing mission to extend its life sufficiently, Hubble and Keck would, in addition to searching for atmospheric dynamics, produce data that could assist in developing a better understanding of Uranus's interior. Such data could help explain Uranus's complex magnetic field without getting direct measurements of the field characteristics, as would be possible only by sending another spacecraft to sample the field close to the planet.

## Context

Every 176 years, planetary alignments are such that it is possible through ingenuous use of gravity-assist (or "slingshot") maneuvers to send a pair of spacecraft to visit all the outer planets from Jupiter to Pluto. Prior to the authorization for Mariner Jupiter-Saturn, which later was named Voyager, the National Aeronautics and Space Administration (NASA) had originally proposed sending a pair of sophisticated spacecraft on a journey that had been called the Grand Tour, to take advantage of this rare opportunity to reach four planets. However, that ambitious plan was not funded. The end result was that Voyager 2 would be the only spacecraft to visit Uranus and Neptune in the twentieth century, and most likely for many decades to come after that initial spacecraft encounter. As such, Voyager 2 has provided the greatest share of data about the Uranian system.

The interpretation of those data led to our current understanding about this unique and in many ways bizarre planet. Despite the tremendous insights provided by the Voyager 2 data, many questions remain unanswered. Among those are important questions concerning the internal structure of the planet, the production of the planet's magnetic field, and the nature of the relatively minor amount of internal heating in the planet.

Just as the Galileo probe orbited Jupiter for a prolonged period of time and the Cassini spacecraft did the same at Saturn, the next logical step for Uranus studies would be to dispatch a dedicated orbiter, a spacecraft outfitted with a wide-ranging suite of scientific instruments to allow focused investigations of the planet's atmosphere, internal structure, magnetic field, ring system, and collection of satellites. However, nearly three decades after the Voyager 2 encounter no such program was on the horizon and no proposal was considered likely to be funded unless a nuclear propulsion capability was developed to lessen the travel time to a planet as far away from Earth as Uranus. In the meantime, Uranus will continue to be studied using the Hubble Space Telescope and ground-based facilities (such as the Keck telescope) in attempts to gain further insight into the many unanswered questions remaining from the Voyager encounter.

*David G. Fisher*

## Further Reading

Bredeson, Carmen. *NASA Planetary Spacecraft: Galileo, Magellan, Pathfinder, and Voyager*. New York: Enslow, 2000. A part of Enslow's Countdown to Space series, this volume provides an overview of NASA planetary exploration during the last two decades of the twentieth century. Designed for younger readers, but suitable for all audiences.

Burgess, Eric. *Uranus and Neptune: The Distant Giants*. New York: Columbia University Press, 1988. Covers the Voyager 2 spacecraft's mission, technical difficulties, and encounters with Jupiter, Saturn, Uranus, and Neptune. Describes data collected by Voyager 2 about Uranus. Includes several illustrations and tables. Well written and suitable for a general audience.

Elkins-Tanton, Linda T. *Uranus, Neptune, Pluto, and the Outer Solar System*. New York: Chelsea House, 2006. This book explores the Sun's relationship with the three outer planets and their moons. It looks at these planets as recorders of the formation of the solar system. Aimed at a general or high school audience. Illustrations, bibliography, index.

Encrenaz, Thérèse, et al. *The Solar System*. New York: Springer, 2004. A thorough exploration of the solar system from early telescopic observations through the space missions that had investigated all planets by the publication date. The astrophysical approach gives our solar system a wider context as just one member of similar systems throughout the universe.

Freedman, Roger A., and William J. Kaufmann III. *Universe*. 8th ed. New York: W. H. Freeman, 2008. A college text on astronomy, somewhat more advanced than many introductory texts, but with a wealth of detail and excellent diagrams. Chapters 6 through 16 describe the solar system. Comes with a CD-ROM.

Hunt, Garry E., and Patrick Moore. *Atlas of Uranus*. New York: Cambridge University Press, 1988. This was the first volume after the 1986 Voyager encounter to offer a comprehensive history of Uranus: its discovery, satellites, rings, and the data returned by Voyager, including photographs.

Irwin, Patrick G. J. *Giant Planets of Our Solar System: An Introduction*. 2d ed. New York: Springer, 2006. Suitable as a textbook for upper-level college courses in planetary science. Focuses on Jupiter, Saturn, Uranus, and Neptune and their satellites, rings, and magnetic fields. Filled with figures and photographs. Accessible to the serious general audience.

Loewen, Nancy. *The Sideways Planet: Uranus*. Mankato, Minn.: Picture Window Books, 2008. An educational children's book devoted to the planet Uranus. Covers the planet's rings, moons, and tilted axis.

McBride, Neil, and Iain Gilmour, eds. *An Introduction to the Solar System*. Cambridge, England: Cambridge University Press, 2004. A complete description of solar system astronomy suitable for an introductory college course but useful to interested laypersons as well. Filled with supplemental learning aids and solved student exercises. A Web site is available for educator support.

Miner, Ellis. *Uranus: The Planet, Rings, and Satellites*. New York: Ellis Horwood, 1990. The author thoroughly covers the topics of both the Uranian system and the Voyager mission. Illustrations, bibliography, index.

Schmude, Richard W. *Uranus, Neptune, and Pluto and How to Observe Them*. New York: Springer, 2008. Ideal for backyard or amateur astronomers who are interested in observing the outer planets. Includes up-to-date information about the planets.

Tocci, Salvadore. *A Look at Uranus*. New York: Franklin Watts, 2003. As part of the Out of This World series, this book covers all aspects of the planet Uranus from its discovery through 2002. Includes several photographs. Suitable for all readers.

# URANUS'S RINGS

**Categories:** Planets and Planetology; The Uranian System

*Uranus has thirteen known rings, eleven inner and two outer ones. These rings are mostly very narrow and faint. Uranus's ring system is less complex than Saturn's but more so than that of Jupiter.*

## Overview

When Uranus was first observed (perhaps as early as 1690), many astronomers considered it to be a star rather than a planet. William Herschel studied Uranus in 1781, when he thought he had discovered a new comet. After two years of further study, astronomers agreed that Uranus was in fact a planet.

Herschel appears to have observed the rings of Uranus in February of 1789. He sketched an image of Uranus in his journal, making a note that the planet had rings of a faint reddish hue. Rings around Uranus remained an open issue for a long time. The existence of Uranus's ring system was finally confirmed, albeit accidentally, in 1977. Astronomers James Elliot, Edward Dunham, and Douglas Mink set out to study Uranus's atmosphere by observing the occultation of the star SAO 158687. They noticed that this star briefly disappeared from view before and after being eclipsed by the planet. Each of the five occultations that they observed yielded the same results. When the group published their work, they referenced these occultations and resulting rings using the Greek letters Alpha, Beta, Gamma, Delta, and Epsilon. In 1978 another group of scientists found four additional rings. The Eta ring was found between the Beta and Gamma rings. The other three were discovered inside the orbit of the Alpha ring and were named Six, Five, and Four (in that order).

The National Aeronautics and Space Administration's (NASA's) robotic Voyager 2 probe flew by Uranus in 1986. The spacecraft took the first photographs of the Uranian ring system. Voyager 2 also discovered two more faint rings, Lambda and 1986U2R/Zeta, bringing the total number of known rings around the planet to eleven. In 2003, the Hubble Space Telescope discovered, and in 2005 confirmed, the existence of an additional pair of rings. They form an outer ring system that is separate from the other eleven rings. These two outer rings have an orbital radius of more than 100,000 kilometers from Uranus's center—double that of the inner rings. In addition to the twelfth and thirteenth rings, Hubble found two satellites. Mab, which is only 24 kilometers in diameter,

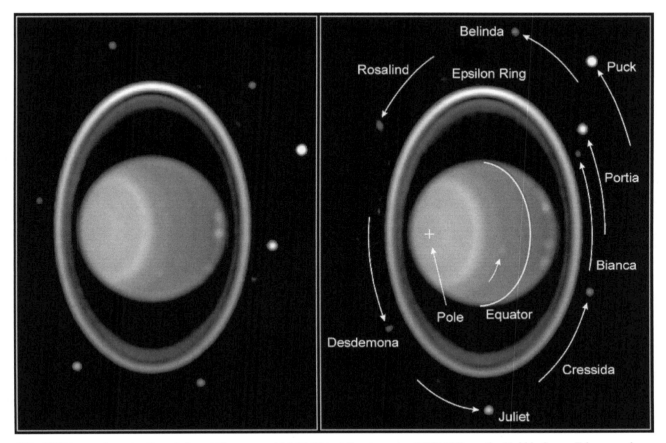

*This 1997 image from the Near Infrared Camera and Multi-Object Spectrometer (NICMOS) on the Hubble Space Telescope shows clouds on Uranus (left) and clearly displays the planet's rings, satellites, and rotation (right). (NASA)*

shares an orbit with the outermost ring. Every time a meteoroid impacts the small satellite, dust particles and other debris ejected become part of the Mu ring. The Nu ring lies between Uranus's small satellites Rosalind and Portia.

The Mu and Nu rings are very different from the inner rings. With widths of 17,000 kilometers and 30,000 kilometers, respectively, the Mu and Nu rings are much broader. These two rings are also much fainter then the others, but they can be seen on the Voyager 2 photographs.

Many similarities exist between Uranus's outer rings and Saturn's E and G rings. The E ring includes Enceladus, which contributes dust to it the same way Mab is believed to contribute to the Mu ring. The Nu ring, like Saturn's G ring, contains no embedded "shepherding" satellites and is composed of dust and larger particles. Scientists working with the Keck telescopes in Hawaii studied the rings at near-infrared wavelengths. The Nu ring was visible, meaning it has a reddish hue. This possibly gives

some credit to Herschel's original claim about observing Uranus's rings, despite critics' claims that the rings are too faint for him to have seen. The Mu ring was not visible, meaning that its small dust particles appear blue in color. Red is a typical color for planetary rings. Blue however, is not. Saturn's E ring is the only other ring known to have the unusual blue hue.

The inner ring system contains two types of rings: narrow and dusty. The closest ring to Uranus is 1986U2R. It was discovered in 1986 by Voyager 2. This ring is only about 12,000 kilometers above the cloud tops of Uranus. The 1986U2R (or Zeta) ring was observed in 2003 and 2004 using the Keck telescopes. Scientists found the ring to be broad, very faint, and composed of dust grains.

The next set of rings is Six, Five, and Four, which were named for the occultations that led to their discoveries. They are the faintest of Uranus's narrow main rings.

## Rings of Uranus

| | Radius (km) | Radius/Eq. Radius | Optical Depth | Albedo (×10⁻3) | Width (km) | Eccentricity |
|---|---|---|---|---|---|---|
| Uranus equator | 25,559 | 1.000 | — | — | — | — |
| 6 | 41,837 | 1.637 | ~0.3 | ~15 | 1.5 | 0.0010 |
| 5 | 42,234 | 1.652 | ~0.5 | ~15 | ~2 | 0.0019 |
| 4 | 42,571 | 1.666 | ~0.3 | ~15 | ~2 | 0.0011 |
| Alpha | 44,718 | 1.750 | ~0.4 | ~15 | 4-10 | 0.0008 |
| Beta | 45,661 | 1.786 | ~0.3 | ~15 | 5-11 | 0.0004 |
| Eta | 47,176 | 1.834 | ~0.4- | ~15 | 1.6 | — |
| Gamma | 47,627 | 1.863 | ~1.3+ | ~15 | 1-4 | 0.0011 |
| Delta | 48,300 | 1.900 | ~0.5 | ~15 | 3-7 | 0.00004 |
| Lambda | 50,024 | 1.957 | ~0.1 | ~15 | ~2 | 0. |
| Epsilon | 51,149 | 2.006 | 0.5-2.3 | ~18 | 20-96 | 0.0079 |

*Source:* Data are from the National Aeronautics and Space Administration/Goddard Space Flight Center, National Space Science Data Center.

These three lie outside Uranus's equatorial plane by 0.06°, 0.05°, and 0.03°, respectively. Six, Five, and Four do not contain dust and are the thinnest of Uranus's narrow rings.

After the Epsilon ring, the Alpha and Beta rings are the brightest of Uranus's rings. Alpha and Beta are narrowest and faintest at their closest points to Uranus. At their farthest, the two rings are their broadest and brightest. Like all of Uranus's rings, Alpha and Beta are composed of extremely dark material. They are much darker than Uranus's inner satellites, meaning that the rings cannot be composed of pure water ice. The composition of the rings is thus unknown, but astronomers think it is probably a mixture of dark materials and ices.

The seventh ring outward from Uranus's core is Eta, at 47,176 kilometers. Eta has both a narrow part and a broader, dustier section. When Voyager 2 photographed the ring in forward scattered light, it appeared very bright, indicating a large amount of dust. Eta has an inclination and eccentricity of zero, meaning that the ring lies in the planet's equatorial plane, and the ring particles execute circular orbits. In 2007 Uranus's rings were viewed edge-on for the first time. The Eta ring appeared to be the second brightest, which was a significant increase. This finding has led planetary scientists to believe that, while the ring is optically narrow, it is geometrically thick. The next chance for astronomers to view the rings in that unique geometry will occur in 2049.

The Gamma ring is narrow, with an inclination close to zero. Gamma's width varies from 3.6 to 4.7 kilometers. The ring was not visible during the 2007 ring plane crossing. This means that Gamma is both optically and geometrically narrow. The ring also does not contain any dust. Scientists are uncertain about what holds this small ring together.

Like the Eta ring, Delta has both a narrow and a broad component. The thinner part varies from 4.1 to 6.1 kilometers wide, while the thicker part ranges from 10 to 12 kilometers. The wide section is composed of dust, unlike the narrow part. In 2007, only the broad area of Delta was visible. At the outer edge of the Delta ring, a small satellite named Cordelia orbits Uranus.

The Lambda ring lies between Cordelia (a shepherd satellite) and the Epsilon ring. Lambda is faint and narrow even when backlit. The dusty ring was first detected by Voyager 2 during stellar occultation observations, but only at ultraviolet wavelengths. The Lambda ring is composed of micrometer-sized dust, which was confirmed in 2007, when it appeared very bright.

The brightest and densest of Uranus's rings is Epsilon. It is the outermost ring of the inner system. Epsilon reflects two-thirds of all the light visible from Uranus's rings. It is only one of two rings that Voyager 2 was able to photograph clearly. Epsilon is the most eccentric of the rings but has a near-zero orbital inclination. The dense ring has particles ranging in size from 0.2 to 20 meters in diameter. In 2007, the ring was not observable, because of its lack of dust. Epsilon contains many dense, narrow ringlets, and possibly partial arcs. The ring may stay so compact because of its shepherd satellites: Cordelia on the inner side and Ophelia on the outer.

## Knowledge Gained

The first rings of Uranus were officially discovered by accident in 1977. Elliot, Dunham, and Mink were using the Kuiper Airborne Observatory (KAO) to study Uranus's atmosphere during five stellar occultations. The KAO is an airplane with a 36-inch (91.5-centimeter) telescope mounted on the side. With the KAO, scientists can conduct research while flying 14 kilometers above the Earth's surface, thereby making infrared observations readily. At that elevation there is significantly less atmospheric water vapor, which blocks infrared wavelengths from reaching the surface of the Earth. The KAO therefore combines many benefits of a space telescope with the accessibility of a ground-based telescope.

Launched in 1977, Voyager 2 is the only spacecraft to have visited Uranus. It came within 81,500 kilometers of the planet on January 24, 1986. The spacecraft had several instruments on board, including cameras, magnetometers, and spectroscopes. At the time, Uranus's south pole was pointed toward the Sun. Voyager discovered ten satellites as well as the Lambda and Zeta (1986U2R) rings. The Mu and Nu rings have since been located on Voyager 2 photographs.

The Hubble Space Telescope was studying Uranus in 2003 when it discovered the Mu and Nu rings. Scientists were able to confirm the finding in 2005. When the Keck telescope studied the two rings at near-infrared wavelengths, only the Nu ring was visible. This means that the Nu ring has a reddish color. The Mu ring therefore has a bluish tint, because it was not visible.

In 2007, astronomers were able to view Uranus's rings edge-on. Teams of scientists used the Keck II telescope in Hawaii, the Hubble Space Telescope, and the European Southern Observatory's Very Large Telescope in Chile to study the event. Images taken with the Keck telescope show that the rings have changed since Voyager 2 visited the planet more than two decades ago. The broad, dusty inner Zeta ring appears very different. If it is the same ring discovered by Voyager, Zeta has moved several thousand kilometers away from Uranus. Similar shifts have been detected in the ring systems of Saturn and Neptune.

## Context

All of the Jovian planets in our solar system have ring systems. Each set of rings is unique. Neptune's rings are simpler than Uranus's, containing only five rings and partial arcs. Saturn's ring system on the other hand is more complex. Jupiter has but two faint rings.

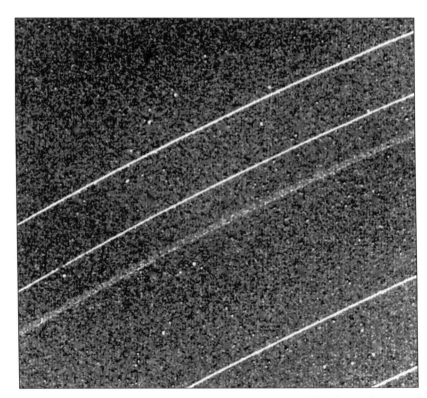

*Voyager 2 took these images of Uranus's rings in January, 1986, from a distance of 1.12 million kilometers.* From the top, the Delta, Gamma, Eta, Beta, and Alpha rings are visible. (NASA/JPL)

The thirteen rings of Uranus are mostly faint and narrow. They are in two groupings—an inner system of eleven rings and an outer set of two. In 2007, the Nu ring was determined to be red in color, and the Mu ring was found to be blue. Red seems to be a typical color for planetary rings, like Saturn's G ring. The blue color of the Mu ring, however, is not common. The only other example in the solar system is Saturn's E ring. What caused this odd blue color is still a mystery to scientists. The rings around Uranus are believed to be made of debris from collisions between Uranus's satellites.

Scientists can learn more about the formation and evolution of the solar system by investigating planetary ring systems. As ground-based and space telescopes improve, astronomers could unlock the secrets of Uranus's rings and the solar system itself.

*Jennifer L. Campbell*

**Further Reading**

Chaisson, Eric, and Steve McMillan. *Astronomy Today*. 6th ed. New York: Addison-Wesley, 2008. A well-written college-level text for introductory astronomy courses. Has a chapter on Uranus and Neptune that covers the ring systems.

Elkins-Tanton, Linda T. *Uranus, Neptune, Pluto, and the Outer Solar System*. New York: Chelsea House, 2006. Explores the Sun's relationship with the three outer planets and their moons, considering these planets as recorders of the formation of the solar system. Aimed at a general or high school audience. Illustrations, bibliography, index.

Esposito, Larry. *Planetary Rings*. New York: Cambridge University Press, 2006. A synopsis of current knowledge of the outer planets' ring systems. Includes information from the Cassini mission on Uranus's rings, and on ring ages and evolution. Geared toward scientists and college students.

Fridman, Alexei M., and Nikolai N. Gorkavyi. *Physics of Planetary Rings: Celestial Mechanics of Continuous Media*. New York: Springer, 1999. Compares the ring systems of Jupiter, Saturn, Uranus, and Neptune using observational and mathematical data. Designed for scientists, astronomy and physics students, and amateur astronomers wishing to know more about the rings of the outer planets.

Hunt, Garry E., and Patrick Moore. *Atlas of Uranus*. New York: Cambridge University Press, 1988. This was the first volume after the 1986 Voyager encounter to offer a comprehensive history of Uranus: its discovery, satellites, rings, and the data returned by Voyager, including photographs.

Loewen, Nancy. *The Sideways Planet: Uranus*. Mankato, Minn.: Picture Window Books, 2008. An educational children's book devoted to the planet Uranus. Covers Uranus's rings, moons, and tilted axis.

Miner, Ellis. *Uranus: The Planet, Rings, and Satellites*. New York: Ellis Horwood, 1990. The author thoroughly covers the topics of both the Uranian system and the Voyager mission. Illustrations, bibliography, index.

Miner, Ellis D., Randii R. Wessen, and Jeffrey N. Cuzzi. *Planetary Ring Systems*. New York: Springer Praxis, 2006. Looks at the ring systems of each gas giant. Covers recent research in the field, as well as the many questions that remain unanswered.

Schmude, Richard W. *Uranus, Neptune, and Pluto and How to Observe Them*. New York: Springer, 2008. Ideal for backyard or amateur astronomers who are interested in observing the outer planets. Includes up-to-date information about the planets.

Tocci, Salvadore. *A Look at Uranus*. New York: Franklin Watts, 2003. As part of the Out of This World series, this book covers all aspects of the planet Uranus, including the planet's discovery and research to 2002. Includes several photographs. Suitable for all readers.

# URANUS'S SATELLITES

**Categories:** Natural Planetary Satellites; Planets and Planetology; The Uranian System

*Uranus's natural satellites form a miniature solar system with distinctive properties that teach us the complexity and diversity of planetary and satellite formation. The peculiar surface features of some of the satellites and their unusual orbital characteristics suggest an earlier epic in the solar system with violent collisions among its members.*

## Overview

Uranus holds its place in the solar system as a member of the subgroup of planets that are called Jovian, after the largest planet in the group, Jupiter. These planets are also referred to as "gas giants" in that their atmospheres comprise the greatest portion of the planet's structure. It is the physical nature and orbital properties of the satellites that set apart one Jovian planet from another, and Uranus is no exception. Surveying the most interesting properties of this planetary system and highlighting their respective features can provide insights about the origin and evolution of the solar system itself.

Uranus has twenty-seven satellites that have been identified. Their names follow a theme that is distinctive in the solar system in that the satellites are not named for mythological figures, like those of the other planets, but instead take their names from characters in plays by William Shakespeare and poems by Alexander Pope. Oberon, Titania, and Puck, for example, were named for characters from Shakespeare's *A Midsummer Night's Dream*; Ariel, Umbriel, and Belinda are named for characters in Pope's *The Rape of the Lock*. Oberon and Titania were first discovered by William Herschel in 1781. Ariel and Umbriel were discovered in 1851 by William Lassell. In 1948, Gerard Kuiper discovered the last moon of any significant size, Miranda.

The only spacecraft to visit Uranus to date has been Voyager 2, which flew by the planet in 1986. Despite the briefness of its visit, Voyager 2 discovered ten small satellites: Juliet, Puck, Cordelia, Ophelia, Bianca, Desdemona, Portia, Rosalind, Cressida, and Belinda. (Perdita was also imaged by Voyager 2, but its discovery was not confirmed until 1999.) The number of satellites has swelled to the current known number of twenty-seven through observation with both orbiting telescopes (Hubble Space Telescope) and ground-based observatories.

A way to organize the Uranian systems of satellites is to think of them as distributed in three divisions. The first division consists of the inner thirteen, relatively small, circular satellites starting just outside the ring system. The second division comprises the five midsized satellites that were discovered prior to the Voyager 2 mission. Finally, the last division comprises nine irregular satellites discovered more recently.

None of the natural satellites of Uranus can be considered on the scale of the largest satellites of the other planets of the solar system, such as Titan (Saturn), Triton (Neptune), the

four Galilean satellites of Jupiter (Io, Europa, Callisto, and Ganymede), or even Earth's own Moon. Nor does any of these satellites contain an atmosphere, as does as Titan, or show active volcanoes, as seen on Io. However, the five largest Uranian satellites in Division 2—Miranda, Ariel, Umbriel, Titania, and Oberon (in order outward from the planet)—all have enough mass to be spherical. Hence, if they were not orbiting Uranus and were free from any debris, they would qualify as dwarf planets, like Pluto. The composition of these satellites is mostly ice, with mixtures of ammonia and methane. They all exhibit synchronous rotation, rotating exactly at the same rate they revolve around Uranus, always showing the same side facing the planet.

The large satellites of Uranus all show extensive cratering. The consensus of scientific opinion is that the larger craters were formed from collisions with planet-sized objects during the formation of the solar system, and the smaller craters were produced afterward from impacts from comets and meteoroids. Ariel, Umbriel, and Miranda have unusual surface features. Ariel, the

*A near-infrared image of the moons of Uranus from the European Southern Observatory, 2002.* (European Southern Observatory)

## Major Uranian Satellites

URANUS    Miranda  Ariel  Umbriel        Titania        Oberon

| Satellite | Diameter (km) | Distance from Planet (km) |
|-----------|---------------|---------------------------|
| Miranda | 500 ± 220 | 130,000 |
| Ariel | 1,300 ± 130 | 192,000 |
| Umbriel | 1,110 ± 100 | 267,000 |
| Titania | 1,600 ± 120 | 438,000 |
| Oberon | 1,630 ± 140 | 596,000 |

*Source:* Data are from Jet Propulsion Laboratory, California Institute of Technology. Voyager at Uranus: 1986. JPL 400-268. Pasadena, Calif.: Author, 1985, p. 5.

lightest-colored, has craters, valleys, and canyons. It has a diameter of about 1,300 kilometers with grooves and crevices that extend over its entire surface, suggesting recent lava flow that has cooled and solidified. Umbriel is the darkest-colored satellite; its surface is old and cratered. Although Umbriel (with a diameter of 1,110 kilometers) shows the most uniform cratering, with little evidence of geological activity, it has a large, bright ring at the top edge of its southern hemisphere; this ring's origin is unknown, but it has unofficially been named the "fluorescent Cheerio." Miranda (about 500 kilometers diameter), the smallest of these five large satellites, has the most distinctive surface features. There are sharp grooves and ridges along its surface. One feature resembles a large chevron, and another looks like a carved racetrack. Titania and Oberon are both larger than the other satellites (with diameters of approximately 1,500-1,600 kilometers), and they are also several times farther away from Uranus. Both Titania and Oberon show large craters, but the presence of circular regions on Oberon suggests that it has experienced more geological activity than Titania. Oberon's surface is frozen and has a mountain 6 kilometers high. Titania, the biggest satellite among those of Uranus, has impact basins, craters, and rifts.

Inside Miranda's orbit there are the thirteen Division 1 satellites, all with enough mass to be spherical and with diameters ranging from about 25 to 170 kilometers. They are (in order out from the planet) Cordelia, Ophelia, Bianca, Cressida, Desdemona, Juliet, Portia, Rosalind, Cupid, Belinda, Puck, and Mab. Their composition appears to be about half water and half rocklike materials. Their orbital distribution presents a more crowded arrangement than that of the large satellites, and it appears that these thirteen bodies interact gravitationally with each other, crossing paths and periodically colliding with one another. Some of these inner satellites may serve as shepherds for Uranus's narrow rings, which orbit closest to the planet.

Outside the orbit of Oberon are the small irregular satellites in Division 3: Francisco, Caliban, Stephano, Trinculo, Sycroax, Margaret, Prospero, Setebos, and Ferdinand. They are considered irregular, though, because they orbit in odd directions far from Uranus. Their respective sizes and compositions have not been measured accurately. Scientists have estimated that their diameters range from 18 to 150 kilometers. They have very eccentric (elliptical) orbits, and all orbit retrograde, that is, in the direction opposite to the general revolution of other bodies in the solar system, which is counterclockwise around the north polar axis of the Sun.

## Knowledge Gained

The study of Uranus's system of natural satellites, including their orbital and physical properties, makes it possible to understand more fully the processes not only of planet formation but of the evolution of the solar system itself. Satellites can be formed at the time of the formation of the planet they orbit, but they can also be captured by the planet's gravitational pull at a later stage. It appears that both processes are at work in the Uranian system. The group of satellites that are in Division 2 (Miranda, Ariel, Umbriel, Titania and Oberon), as well as the thirteen satellites in Division 1, most likely formed at the same time Uranus condensed. However, events have transpired to alter these satellites' orbits and surface features. For example, Miranda has a scarred surface that is presumed to be the result of a collision with another object and was fractured into pieces. These pieces then came back together unevenly and show the rough terrain and scarps that characterize the satellite. An alternative hypothesis is that Miranda, being too small an object to complete its internal mixing, froze midway through the process of separating its structure into layers.

Several of the Division 1 moons are in such close orbits with each other that they conceivably collide and switch orbital positions around Uranus. In addition, Cordelia and Ophelia serve as shepherding moons for the Epsilon ring system.

The irregular satellites in Division 3 appear to be satellites that were previously solar system bodies, such as comets and asteroids, that were captured by Uranus's gravitational pull and orbit around the planet in elliptical and retrograde orbits.

This accumulated knowledge points to a solar system that evolved sequentially over vast periods of time with many diverse objects that are still changing their orbital shapes and constitutions.

## Context

There were only five known satellites of Uranus until Voyager 2 visited this planetary system and made its closet approach on January 24, 1986. Voyager 2 was launched in 1977 and visited Jupiter in 1979, Saturn in 1981, and, after its flyby of Uranus, Neptune in 1989.

As Voyager 2 approached the Uranus system, its onboard computers were reprogrammed by scientists and engineers back on Earth at the Jet Propulsion Lab to enable the cameras to produce high-quality photographs in the reduced light and at the high speeds at which the spacecraft would be traveling on its flyby. Most of the photographs were taken in a six-hour period in and around the time of closest approach (9:59 A.M. PST) on January 24. On the way to the rendezvous with Uranus, Voyager 2 obtained clear, high-resolution images of each of the five large Uranian satellites (Miranda, Ariel, Umbriel, Titania, and Oberon). It was its discovery of eleven new satellites that added to our knowledge base of this planet. During processing of images of the outer ring (Epsilon ring) of Uranus, it was discovered that two small satellites, Cordelia and Ophelia, were shepherding or keeping this thin ring in orbit around the planet. Also sighted were Bianca, Cressida, Desdemona, Juliet, Portia, Rosalind, Cupid, Belinda, and Perdita which belong to what has been called the Portia Group of satellites, These satellites have similar orbits and light-reflecting properties. The closeness of their respective orbits leads to the hypothesis that this group interacts with each other and may at times collide.

Until 1997, the Uranian system was distinct from the other Jovian planets in that there were no identified irregular satellites. However, the discovery on September 6, 1997, of two irregular satellites, Caliban and Sycroax, by Brett J. Gladman, Philip D. Nicholson, Joseph A. Burns, J. J. Kavelaars, Brian G. Marsden, Gareth V. Williams, and Warren B. Offutt, using the 200-inch Hale telescope (at Palomar Observatory in Southern California) removed that distinction. Subsequently, Stephano, Prospero, and Setebos were discovered by Matthew J. Holman, Kavelaars, Gladman, Jean-Marc Petit, and Hans Scholl on July 18, 1999. Trinculo, Margaret, and Ferdinand were discovered by Holman, Kavelaars, and Dan Milisavljevic on August 13, 2001.

*Joseph Di Rienzi*

*Light-colored Ariel, seen in this 1986 image from Voyager 2, has grooves and crevices that extend over its entire surface, suggesting recent lava flow that has cooled and solidified.* (NASA/JPL)

## Further Reading

Bennett, Jeffery, Megan Donahue, Nicholas Schneider, and Mark Voit. *The Cosmic Perspective*. 3d ed. San Francisco: Pearson Addison Wesley, 2004.

This textbook provides a thematically organized overview of the universe. Chapter 12 discusses the Jovian systems and contains a subsection on the medium-sized satellites of Uranus.

Burgess, Eric. *Uranus and Neptune: The Distant Giants*. New York: Columbia University Press, 1988. Covers the Voyager 2 spacecraft's mission, technical difficulties, and its encounters with Jupiter, Saturn, and Uranus. Describes the data collected by Voyager 2 about Uranus. Includes several illustrations and tables. Well written and suitable for the general audience.

Carroll, Bradley W., and Dale A. Ostlie. *An Introduction to Modern Astrophysics*. San Francisco: Pearson Addison Wesley, 2007. This is an encyclopedic textbook that covers all of modern astronomy and astrophysics. Although much of the book is for the advanced student, the chapters on the solar system are very descriptive. Chapter 21, "The Realms of the Giant Planets," includes a section on their satellites and discusses Miranda in particular.

Elkins-Tanton, Linda T. *Uranus, Neptune, Pluto, and the Outer Solar System*. New York: Chelsea House, 2006. This book explores the Sun's relationship with the three outer planets and their moons. It looks at these planets as recorders of the formation of the solar system. Aimed at a general or high school audience. Illustrations, bibliography, index.

Encrenaz, Thérèse, et al., eds. *The Outer Planets and Their Moons: Comparative Studies of the Outer Planets Prior to the Exploration of the Saturn System by Cassini-Huygens*. New York: Springer, 2005. An in-depth look at the current understanding of the solar system's outer planets. Focuses on the studies of their formation, evolution, magnetospheres, satellites, and ring structures. For scientists, first-year graduate students, and advanced undergraduates.

Hunt, Garry E., and Patrick Moore. *Atlas of Uranus*. New York: Cambridge University Press, 1988. This was the first volume after the 1986 Voyager encounter to offer a comprehensive history of Uranus: its discovery, satellites, rings, and the data returned by Voyager, including photographs.

Loewen, Nancy. *The Sideways Planet: Uranus*. Mankato, Minn.: Picture Window Books, 2008. An educational children's book devoted to the planet Uranus. Covers Uranus's rings, moons, and tilted axis.

Miner, Ellis. *Uranus: The Planet, Rings, and Satellites*. New York: Ellis Horwood, 1990. The author thoroughly covers the topics of both the Uranian system and the Voyager mission. Miranda is featured as a remarkable satellite. Illustrations, bibliography, index.

Schmude, Richard W. *Uranus, Neptune, and Pluto and How to Observe Them*. New York: Springer, 2008. Ideal for backyard or amateur astronomers who are interested in observing the outer planets. Also includes up-to-date information about the planets.

Tocci, Salvadore. *A Look at Uranus*. New York: Franklin Watts, 2003. As part of the Out of This World series, this book covers all aspects of the planet Uranus, from its discovery through 2002. Includes several photographs. Suitable for all readers.

# URANUS'S TILT

**Categories:** Planets and Planetology; The Uranian System

*All planets in our solar system rotate on an axis tilted in relation to the ecliptic plane (the plane carved out by their orbit around the Sun). However, Uranus's axis is tilted at such an extreme angle that the planet rotates while virtually lying on its side. Despite several theories, scientists do not fully understand what causes Uranus's tilt.*

### Overview

Uranus was observed as early as 1690, but astronomers thought it was a star. Using a telescope he built himself, Sir William Herschel observed Uranus over a series of nights in 1781. Herschel initially reported to the Royal Society that he had discovered a new comet. After tracking the "comet" for two years, astronomers finally agreed that Uranus was actually the seventh planet in the solar system.

In 1829, astronomers determined that the rotation of Uranus was unique. All of the planets rotate on an axis that is tilted with respect to the orbital plane of the solar system. The orbital plane is the imaginary surface on which the planets orbit and almost lies on the Sun's equator (the plane is tilted at a 7° angle with respect to the Sun's equator). Axial tilt is calculated by drawing a line perpendicular to the orbital plane. The rotational axis of the planet is compared to the perpendicular line. For example, the rotational axis of the Earth has a tilt of 23.5°. Mars is tilted at 25.19°, and Saturn's axis is tilted at 26.73°. Uranus, on the other hand, has an axial tilt of 97.8°. Because of this, Uranus is often referred to as the "sideways planet." Either the north or the south pole of Uranus is usually pointed toward the Sun. Uranus's equator experiences day and night the same way as the Earth's polar ice caps do. The poles of Uranus each experience forty-two years of sunlight, followed by forty-two years of complete darkness. Only around equinoxes is the Sun facing Uranus's equator, causing "normal" Earth day-night conditions. Its last equinox occurred on December 7, 2007, and the the next will not happen until the year 2049.

Due to its unusual orientation, scientists have conflicting methods for determining which pole is "north" and which is "south." The International Astronomical Union (IAU) refers to whichever pole lies above the orbital plane as the north pole. Most scientists use this designation. Others use the right-hand rule from physics and the direction the planet is spinning to designate the poles north or south. This method contradicts the IAU's determination, instead naming the pole below the orbital plane as "north."

The only spacecraft to date that has visited Uranus is Voyager 2. Launched in 1977, the probe reached Uranus in 1986. Voyager 2 came within 81,500 kilometers of the planet. It discovered and photographed ten new satellites and nine rings orbiting Uranus. The spacecraft also helped scientists determine more precisely the axial tilt of Uranus.

There are two main competing theories to explain why Uranus is tilted on its side. No one knows who proposed the popular "collision" theory, which posits that Uranus formed and then a large Earth-sized object crashed into it with such force that it left the planet on its side. The current accepted theory of planetary formation is the idea of nebular condensation, developed in the seventeenth century by French philosopher René Descartes. As a massive cloud of interstellar dust and debris condensed, it would collapse and start to spin. Planets slowly would begin to form from clumps of matter joining together. The bigger the planets grew, the faster they would be able to attract more material through a process known as accretion. The debris cloud that the planets formed from is called the accretion disk, which became the orbital plane. The collision knocking Uranus on its side would have had to happen early in its formation. Possibly an object struck the planet's core before Uranus's satellites had condensed from the debris cloud surrounding it. Another theory is that the impact left behind debris that later became Uranus's satellites. However, there are several questions that remain unanswered with this scenario.

Why does Uranus have a nearly circular obit, like the other planets? Would not a large impact have affected Uranus's orbit? If Uranus's satellites had formed before the collision, why were their orbits not changed? The satellites orbit Uranus's equator, just like its ring system. Two very small captured satellites, however, have been found orbiting Uranus's poles. The nebular theory also fails to explain other oddities of the solar system, such as why Venus has a retrograde rotation (rotates backward), why Mercury and Pluto have elliptical orbits, and why Uranus and Neptune have tilted magnetic fields.

In 1997, Argentinean scientists Adrian Brunini and Mirta Parisi published a paper giving plausible ways that Uranus became tilted. They believed that if a collision had taken place, it had to be when Uranus was a more solid core surrounded by a planetary envelope. The impacting object would have hit the proto-Uranus from the opposite direction as it traveled around the Sun. The two scientists thought that studying Uranus's satellites was the key to figuring out its odd axial tilt. They concluded that either the satellites of Uranus were created by the collision itself, or no collision happened. Brunini and Parisi's study found that Uranus's satellite Prospero (S/1999 U3) set a number of constraints on any possible conclusion. Therefore, they believe, it is possible that a new theory of solar-system formation is needed to explain the tilt of Uranus. A number of scientists seem to be shifting toward the second explanation: that Uranus formed tilted on its side, and that the nebular theory for the formation of the solar system fails to explain how this could have happened. Researchers have been working on finding a simulation that solves this and other oddities of the solar system.

In 2006, Brunini published a new theory of the formation of the solar system in *Nature* magazine. His mathematical model is based on the idea that Jupiter and Saturn once had a 1:2 orbital resonance. This means that in the time it took Saturn to orbit the Sun once, Jupiter went around twice. The gravitational effect of Jupiter and Saturn gradually changed the orbits of Uranus and Neptune. Brunini's simulation shows that his model would take about a million years for the outer planets to reach the orbital positions we now observe. He argues that during the close encounter of Saturn and Uranus, the angular momentum of the planets shifted, which over time caused their axial tilts to change. This scenario, Brunini argues, can explain the orbits of Uranus's rings and satellites, which would have slowly changed their orientation along with Uranus. Unlike a collision, Brunini's scenario would have taken hundreds of thousands of years to play out.

No definitive answer has been found for what caused Uranus's unique tilt. Only Voyager 2 has visited Uranus; new spacecraft would be able to provide more data but cannot be sent until either the planets are again aligned for a "slingshot" (gravity-assist) approach (more than a century away) or the necessary nuclear propulsion systems are developed. Until then, researchers are left making mathematical and computer models in their efforts to solve the mysteries of Uranus's axial tilt.

## Methods of Study

Uranus can be observed from Earth with telescopes, and on dark, clear nights can be viewed with the unaided eye. Scientists have also taken photographs of the Uranus system using the Hubble Space Telescope. In late 2002, astronomers in Chile were able to image Uranus, its rings, and some of its satellites. The pictures were taken with the Very Large Telescope (VLT) at the European Southern Observatory (ESO) Paranal Observatory. The rings that are normally unable to be viewed from Earth, along with seven satellites, appeared in the image because it was taken at near-infrared wavelengths.

The Voyager program is the only spacecraft that has visited Uranus. Launched in 1977, Voyager 2 came within 81,500 kilometers of Uranus on January 24, 1986. Voyager 2 was equipped with more than a dozen scientific instruments, including cameras, television cameras, magnetometer, and spectroscopes. Voyager 2 viewed Uranus's "south" pole (located south of the orbital plane), which was pointed toward the Sun. At Uranus, Voyager 2 discovered ten satellites and two rings. The spacecraft also studied the planet's five largest moons (Oberon, Umbriel, Titania, Ariel, and Miranda), taking the first close-up photographs of them. Voyager 2 provided the first close-up photographs of Uranus and detailed information about its magnetic field, ring system, weather, and unusual axial tilt. By the early nineteenth century, scientists knew that the planet was tilted, but it was not until Voyager 2 arrived at Uranus that astronomers knew precisely how tilted it was.

Computer modeling can be used to explore the dynamics of complex systems over time, where geologic time is essentially replaced by computation time. Modern computers allow a tremendous amount of computational power, and the magnitude of that computational power is continuously increasing. Basically, a computer modeling effort such as that used by Brunini and similar researchers seeks to begin with certain basic assumptions about initial conditions of a complex system such as Uranus in its interaction with larger bodies such as Jupiter and Saturn, and then introduce the gravitational interactions between all of these bodies and allow the computational cycle to mimic the passage of time as each of these bodies orbits the Sun and continues to interact with the others. This sort of thing cannot be easily done by hand. Sir Isaac Newton, in presenting his development of mechanics in *Philosophiae Naturalis Principia Mathematica* (1687; commonly known as *The Principia*), provided a means of quantifying the gravitational interaction between two bodies. That relatively simple problem can be solved in closed form in both spatial and temporal coordinates. However, the three-body problem requires numerical analysis, which is largely done at present by computer programming or software packages, since it cannot be solved in closed form. The more bodies are involved in a calculation, the more computing power is required.

## Context

Scientists may never know the real reason for Uranus's axial tilt. Further study of the planet Uranus by spacecraft, and even possibly by humans, could lead to the answer. Computer simulations and mathematical models can help scientists speculate what might have happened. Maybe the accepted nebular theory for how the solar system formed is incorrect. Maybe the gravitational effects of Jupiter and Saturn slowly caused Uranus to lean to its side. Maybe Adrian Brunini and his colleagues are correct, and the only way to make this determination is to study Uranus's satellites. The quest to explain Uranus's extreme axial tilt could lead to a new view of how Earth and the solar system formed.

*Jennifer L. Campbell*

## Further Reading

Burgess, Eric. *Uranus and Neptune: The Distant Giants*. New York: Columbia University Press, 1988. Covers the Voyager 2 spacecraft's mission, technical difficulties, and its encounters with Jupiter, Saturn, and Uranus. Focuses on data collected by Voyager 2 about Uranus. Includes several illustrations and tables. Well written, suitable for the general audience.

Chaisson, Eric, and Steve McMillan. *Astronomy Today*. 6th ed. New York: Addison-Wesley, 2008. A well-written college textbook for introductory astronomy courses. Includes a chapter on Uranus and Neptune.

Elkins-Tanton, Linda T. *Uranus, Neptune, Pluto, and the Outer Solar System*. New York: Chelsea House, 2006. Explores the Sun's relationship with the three outer planets and their moons. Looks at these planets as recorders of the formation of the solar system. Aimed at a general or high school audience. Illustrations, bibliography, index.

Encrenaz, Thérèse, et al., eds. *The Outer Planets and Their Moons: Comparative Studies of the Outer Planets Prior to the Exploration of the Saturn System by Cassini-Huygens*. New York: Springer, 2005. An in-depth look at the current understanding of the solar system's outer planets. Focuses on the studies of their formation, evolution, magnetospheres, satellites, and ring structures. For scientists, first-year graduate students, and advanced undergraduates.

Fraknoi, Andrew, David Morrison, and Sidney Wolff. *Voyages to the Stars and Galaxies*. Belmont, Calif.: Brooks/Cole-Thomson Learning, 2006. An introductory college text that gives students easy-to-understand analogies to help them with more complex theories. Well written and easy to read. Includes a CD-ROM featuring InfoTrac software.

Freedman, Roger A., and William J. Kaufmann III. *Universe*. 8th ed. New York: W. H. Freeman, 2008. A thorough and well-written introductory college astronomy textbook. Covers all aspects of Uranus.

Hunt, Garry E., and Patrick Moore. *Atlas of Uranus*. New York: Cambridge University Press, 1988. This was the first volume after the 1986 Voyager encounter to offer a comprehensive history of Uranus: its discovery, satellites, rings, and the data returned by Voyager, including photographs.

Loewen, Nancy. *The Sideways Planet: Uranus*. Mankato, Minn.: Picture Window Books, 2008. An educational children's book devoted to the planet Uranus. Covers Uranus's rings, moons, and tilted axis.

Miner, Ellis. *Uranus: The Planet, Rings, and Satellites*. New York: Ellis Horwood, 1990. The author thoroughly covers the topics of both the Uranian system and the Voyager mission. Illustrations, bibliography, index.

Schmude, Richard W. *Uranus, Neptune, and Pluto and How to Observe Them*. New York: Springer, 2008. Ideal for backyard or amateur astronomers who are interested in observing the outer planets. Also includes up-to-date information about the planets.

Tocci, Salvadore. *A Look at Uranus*. New York: Franklin Watts, 2003. As part of the Out of This World series, this book covers all aspects of the planet Uranus, from its discovery through 2002. Includes several photographs. Suitable for all readers.

# VOYAGER PROGRAM

**Category:** Space Exploration and Flight

*The Voyager Program probes executed the first "Grand Tour" in planetary exploration by successively encountering the outer planets: Jupiter, Saturn, Uranus, and Neptune. Such a tour, using the "planetary-gravity-assist" technique to travel from planet to planet, is possible only once every 175 years.*

### Overview

The Voyager Program conducted the first planetary Grand Tour in history by sending two uncrewed spacecraft on a mission to explore the outer solar system. Preliminary design began in 1969. After obtaining official approval in May, 1972, Voyager 1 was launched on September 5, 1977. Voyager 2 was actually launched a few weeks earlier on August 20, 1977. Both were launched from Cape Canaveral, Florida, by Titan III-E/Centaur rockets. Voyager 1 encountered Jupiter in 1979 and Saturn in 1980, before flying out of the plane of the solar system. Voyager 2 encountered Jupiter in 1979, Saturn in 1981, Uranus in 1986, and Neptune in 1989. As of 2001, the two Voyager spacecraft are the most distant human-made objects from Earth and have provided humankind's most detailed views of the outer solar system.

The Voyager Program was originally approved as a mission to Jupiter and Saturn only. However, the mission scientists and engineers knew that during the 1970's, the configuration of the outer planets in the solar system provided a unique opportunity for a Grand Tour of the outer solar system. With this Grand Tour in mind, they designed the Voyager craft with the capability of extending the original mission should approval later be granted.

After Voyager's early success, the National Aeronautics and Space Administration (NASA) officially approved the extension of the mission. Voyager 1 turned to fly out of the plane of the solar system after its encounter with Saturn. Voyager 2 continued on to explore Uranus and Neptune. The only possible trajectory that would have allowed Voyager 2 to continue its mission to Pluto went directly through the interior of the planet Neptune.

During each planetary flyby, the Voyager craft used the gravity-assist technique, a sort of gravitational slingshot effect from the giant planets that propelled the spacecraft on to the next planet. The Voyager spacecraft trajectories were very carefully selected so that as the Voyager fell toward a planet, the craft would pick up speed. The planet also deflected the trajectory so that the craft was aimed in the right direction for the next planetary encounter. These gravitational boosts shortened the time required for Voyager 2 to reach Neptune by nearly two decades and significantly reduced the amount of fuel necessary to propel the craft from planet to planet.

**Envoys Beyond the Solar System**
On February 17, 1998, at a distance of 6.5 billion miles, Voyager 1 exceeded the distance from the Sun of the slower Pioneer 10 spacecraft (launched in 1972) to become the most distant human-made craft from Earth. As of January, 2001, the Voyager 1 and 2 craft remained functional and were located more than 7.4 billion and 5.8 billion miles from Earth, respectively. They have enough power to last until about 2020, when they will both be more than 10 billion miles from Earth. During this time, they may cross the heliopause, the boundary between the solar system and the surrounding universe, to become the first human-made objects to leave the solar system. Voyager 1 will pass near a faint star in the constellation Camelopardis in about 40,000 years. In 296,000 years Voyager 2 will pass near the brightest star in the night sky, Sirius.

As humankind's envoys beyond the solar system, both Voyager craft carry records containing pictures and sounds from Earth. The records are 12-inch copper disks plated with gold. The craft also contain needles for playing the records and illustrations of their use. More than one hundred pictures were included, twenty of which are in color. These pictures attempt to depict Earth and its rich variety of life, including human life and culture. The records contain spoken greetings in sixty different languages, natural and machine-made sounds of Earth, and a 90-minute selection of musical excerpts that represent a variety of cultures and forms.

**Instrument Packages**
Each Voyager craft, which is about the size and mass of a subcompact car, contains a suite of scientific instruments. The Voyager images were provided by two cameras on each craft. They have filter wheels to allow color images. The narrow-angle cameras, capable of resolving a newspaper headline from a distance of 1 kilometer, provided high-resolution images. The wide-angle cameras provided the global images at a lower resolution.

The Voyager crafts' infrared and ultraviolet spectrometers and photopolarimeters provided information about atmospheric and satellite compositions and structures. The planetary radio astronomy experiment measured the planetary radio emissions. The magnetometers measured and studied the planetary magnetic fields and their interactions with the solar magnetic field. Four experiments—the plasma-particles experiment, the plasma-waves experiment, the low-energy charged-particles experiment, and the cosmic-ray particles experiment—were designed to provide information about energetic charged particles at the planetary encounters and in interplanetary space.

**Jupiter Encounter**
Voyager 1 flew within 217,000 miles of Jupiter on March 5, 1979. Voyager 2 followed on July 9, 1979, flying within 449,000 miles. The detailed images and measurements taken during the Voyagers' encounter with Jupiter vastly exceeded what had been possible to accomplish from Earth or on previous Pioneer missions and revealed much new information about Jupiter and its system of moons. Pictures of Jupiter revealed beautiful detail in the striped zone-and-belt structure of the planet's cloud tops. The zones are the lighter-colored stripes, and the belts are the darker-colored stripes. The Voyagers showed that these zones and belts are manifestations of both east-west and up-down circulation patterns in the atmosphere. The high wind speeds are manifested by the obvious turbulence in the zone-belt interfaces. The previously known Great Red Spot is a centuries-old anticyclonic storm larger than Earth. It rotates fully counterclockwise within a period of four to six days. The Voyagers also measured a strong planetary magnetic field for Jupiter.

Surrounding Jupiter, the Voyager mission discovered a ring system, albeit much less extensive than that of Saturn, as well as three small, previously undiscovered moons. Both craft observed Jupiter's previously known moons in unprecedented detail, and Voyager 1 made the unexpected observation that Jupiter's closest major moon, Io, had nine active volcanoes at the time of the Voyager 1 encounter. The next major moon, Europa, has an icy crust covered with a large number of intersecting cracks, beneath which there may be a liquid ocean. Voyager found two distinct types of terrain, grooved and cratered, on Jupiter's next major moon, Ganymede. The final major moon, Callisto, has an ancient surface saturated with craters.

### Saturn Encounter

On November 12, 1980, Voyager 1 flew to within 77,000 miles of Saturn, followed by the Voyager 2 on August 25, 1981, which flew to within 63,000 miles of Saturn. Both craft returned unprecedented data. Pictures of Saturn revealed a zone-and-belt structure to the cloud tops similar to that of Jupiter. However, Saturn's zones and belts did not have the richly detailed structure found on Jupiter, despite Saturn's higher measured wind speeds, which can exceed 1,000 miles per hour. Apparently, this zone-and-belt structure is slightly deeper on Saturn than on Jupiter and covered by a hazy layer that masks its detail. The Voyagers also measured Saturn's magnetic field and confirmed that it is the only planetary magnetic field almost perfectly aligned with the rotation axis.

Saturn's most beautiful feature is its extensive ring system, the only one that is directly visible from Earth. Voyager photographs revealed surprisingly detailed and completely unexpected structures. The major rings, called the A-, B-, and C-rings, consist of hundreds of smaller individual rings. They are apparently caused by the combined gravitational forces of Saturn's many moons on the individual ring particles orbiting Saturn. The faint outer F-ring was revealed to consist of apparently twisted strands. Dark spokes were found in the B-ring. Some of these spokes rotate with Saturn's magnetic field rather than at the orbital speed of ring particles.

The Voyager mission discovered six new moons of Saturn and studied the planet's previously known moons. Saturn's largest moon, Titan, was found to have a significant atmosphere, consisting primarily of nitrogen, like Earth's atmosphere, but much colder. Further studies may help scientists understand the chemistry of Earth's primitive atmosphere. With a diameter of only about 300 miles, Enceladus should be too small to be geologically active, yet it surprisingly proved to be the most geologically active of all Saturn's moons, except Titan. The source of this activity is poorly understood. The moon Mimas is slightly less than 250 miles in diameter, but it has a 6-mile deep impact crater that is 80 miles in diameter. This crater also has a central mountain comparable in size to Mount Everest. With this crater, Mimas resembles the Death Star of the motion picture *Star Wars*.

### Uranus and Neptune Encounters

After the Saturn encounter, Voyager 1 exited the plane of the solar system. Voyager 2 continued on its Grand Tour to explore Uranus and Neptune. The challenges of the outer solar system necessitated extensive reprogramming during the four-year voyage from Saturn to Uranus. At a distance of nearly 2 billion miles from the Sun, it is much darker at Uranus. Hence, long time exposures are needed for the images. Because Voyager 2 was speeding by on its outward journey, it was not possible to mount the cameras on a steady tripod, as is usually done for timed exposures. In a difficult maneuver, called image-motion compensation, the craft rotated just the right amount to compensate for its motion. This technique worked to produce clear sharp images of Uranus and Neptune. In addition, with only a 23-watt radio transmitter the method used to transmit data back to Earth had to be revised, and the receiving antennaes on Earth had to be linked to pick up the distant signal.

On January 24, 1986, Voyager 2 flew to within 67,000 miles of Uranus, taking images that revealed a nearly featureless pale blue planet. With extensive computer processing, the images show a barely visible cloud structure on Uranus. Three years later, on August 24-25, 1989, Voyager 2 flew to within only 3,000 miles of Neptune's north pole. In contrast to Uranus, Neptune showed considerable atmospheric activity. There was a dark blue spot called the Great Dark Spot, similar to the Great Red Spot of Jupiter, and a few smaller dark spots. There were also white lenticular-shaped clouds above the Great Dark Spot.

Both Uranus and Neptune have fairly strong magnetic fields, with large tilts relative to their planetary spin axes. The source of these magnetic fields is poorly understood. The Voyager cameras also studied the thin ring systems around both planets. Voyager 2 studied the previously known moons around both planets and discovered ten new moons around Uranus and six around Neptune.

### Context

On February 14, 1990, at a distance of 3.7 billion miles from Earth, Voyager 1 took a unique family portrait of the solar system. A series of 39 wide-angle images shows the Sun and all but the three smallest planets, Mercury, Mars, and Pluto. The narrow-angle camera also took images of the six visible planets. From this perspective, Earth is a faint dot.

The success of the Voyager mission far exceeded expectations. On their twelve-year mission to the outer solar system the Voyager craft sought out and explored many strange new worlds. Following the Pioneer 10 and 11 missions, the Voyagers were not the first missions to Jupiter and Saturn. However, the Voyager mission was the first to distant Uranus and Neptune. The Voyager mission provided the first detailed look at all these planets. These hardy craft will also be the first human-made objects to leave the solar system during the first two decades of the twenty-first century.

*Paul A. Heckert*

## Further Reading

Hartmann, William K. *Moons and Planets*. 3d ed. Belmont, Calif.: Wadsworth, 1993. Written from a comparative planetology perspective, this book has integrated text devoted to the worlds studied by Voyager organized by specific topics.

Morrison, David, and Jane Samz. *Voyage to Jupiter: NASA SP-439*. Washington, D.C.: U.S. Government, 1980. This book describes the Voyager mission in general and provides the details of the Jupiter encounter.

Sagan, Carl, F. D. Drake, Ann Druyan, Timothy Ferris, Jon Lomberg, and Linda Salzman Sagan. *Murmurs of Earth*. New York: Random House, 1978. This book describes extensively the pictures and sounds on the records contained on each of the Voyager craft.

# INDEX